Invariant Means and Finite Representation Theory of C^*-Algebras

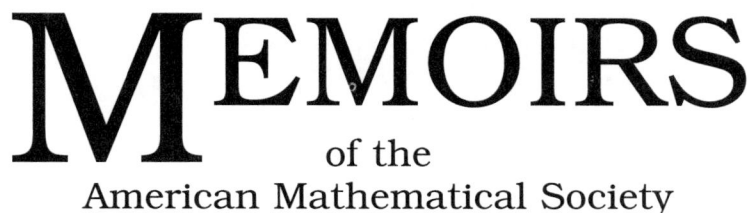

Memoirs
of the
American Mathematical Society

Number 865

Invariant Means and Finite
Representation Theory
of C^*-Algebras

Nathanial P. Brown

November 2006 • Volume 184 • Number 865 (first of 4 numbers) • ISSN 0065-9266

American Mathematical Society
Providence, Rhode Island

2000 *Mathematics Subject Classification.* Primary 46L05.

Library of Congress Cataloging-in-Publication Data
Brown, Nathanial P. (Nathanial Patrick), 1972–
  Invariant means and finite representation theory of C^*-algebras / Nathanial P. Brown.
    p. cm. — (Memoirs of the American Mathematical Society, ISSN 0065-9266 ; no. 865)
  "Volume 184, number 865 (first of 4 numbers)."
  Includes bibliographical references.
  ISBN-13: 978-0-8218-3916-4 (alk. paper)
  1. C^*-algebras. 2. Representation theory. I. Title.

QA326.B76   2006
512′.556—dc22   2006043005

Memoirs of the American Mathematical Society

This journal is devoted entirely to research in pure and applied mathematics.

Subscription information. The 2006 subscription begins with volume 179 and consists of six mailings, each containing one or more numbers. Subscription prices for 2006 are US$624 list, US$499 institutional member. A late charge of 10% of the subscription price will be imposed on orders received from nonmembers after January 1 of the subscription year. Subscribers outside the United States and India must pay a postage surcharge of US$31; subscribers in India must pay a postage surcharge of US$43. Expedited delivery to destinations in North America US$35; elsewhere US$130. Each number may be ordered separately; *please specify number* when ordering an individual number. For prices and titles of recently released numbers, see the New Publications sections of the *Notices of the American Mathematical Society*.

Back number information. For back issues see the *AMS Catalog of Publications*.

Subscriptions and orders should be addressed to the American Mathematical Society, P. O. Box 845904, Boston, MA 02284-5904, USA. *All orders must be accompanied by payment.* Other correspondence should be addressed to 201 Charles Street, Providence, RI 02904-2294, USA.

Copying and reprinting. Individual readers of this publication, and nonprofit libraries acting for them, are permitted to make fair use of the material, such as to copy a chapter for use in teaching or research. Permission is granted to quote brief passages from this publication in reviews, provided the customary acknowledgment of the source is given.

Republication, systematic copying, or multiple reproduction of any material in this publication is permitted only under license from the American Mathematical Society. Requests for such permission should be addressed to the Acquisitions Department, American Mathematical Society, 201 Charles Street, Providence, Rhode Island 02904-2294, USA. Requests can also be made by e-mail to `reprint-permission@ams.org`.

Memoirs of the American Mathematical Society is published bimonthly (each volume consisting usually of more than one number) by the American Mathematical Society at 201 Charles Street, Providence, RI 02904-2294, USA. Periodicals postage paid at Providence, RI. Postmaster: Send address changes to Memoirs, American Mathematical Society, 201 Charles Street, Providence, RI 02904-2294, USA.

© 2006 by the American Mathematical Society. All rights reserved.
This publication is indexed in *Science Citation Index*®, *SciSearch*®, *Research Alert*®, *CompuMath Citation Index*®, *Current Contents*®/*Physical, Chemical & Earth Sciences*.
Printed in the United States of America.

∞ The paper used in this book is acid-free and falls within the guidelines established to ensure permanence and durability.
Visit the AMS home page at `http://www.ams.org/`

10 9 8 7 6 5 4 3 2 1   11 10 09 08 07 06

To my big, beautiful family – on both sides of the Pacific.

Contents

Chapter 1.	Introduction	1
Chapter 2.	Notation, definitions and useful facts	9
Chapter 3.	Amenable traces and stronger approximation properties	15
3.1.	Characterizations of amenable traces	15
3.2.	Uniform amenable traces	22
3.3.	Quasidiagonal traces	25
3.4.	Locally finite dimensional traces	27
3.5.	Miscellaneous remarks and permanence properties	27
Chapter 4.	Examples and special cases	35
4.1.	Discrete groups	35
4.2.	Nuclear and WEP C*-algebras	39
4.3.	Locally reflexive, exact and quasidiagonal C*-algebras	42
4.4.	Type I C*-algebras	45
4.5.	Tracially AF C*-algebras	47
Chapter 5.	Finite representations	55
5.1.	II_1-factor representations of some universal C*-algebras	55
5.2.	Elliott's intertwining argument for II_1-factors	57
5.3.	II_1-factor representations of Popa Algebras	58
Chapter 6.	Applications and connections with other areas	67
6.1.	Elliott's classification program	67
6.2.	Counterexamples to questions of Lin and Popa	77
6.3.	Connes' embedding problem	82
6.4.	Amenable traces and numerical analysis	87
6.5.	Amenable traces and obstructions in K-homology	92
6.6.	Stable finiteness versus quasidiagonality	97
6.7.	Questions	100
Bibliography		103

Abstract

Various subsets of the tracial state space of a unital C*-algebra are studied. The largest of these subsets has a natural interpretation as the space of invariant means. II_1-factor representations of a class of C*-algebras considered by Sorin Popa are also studied. These algebras are shown to have an unexpected variety of II_1-factor representations. In addition to developing some general theory we also show that these ideas are related to numerous other problems in operator algebras.

Received by the editor May 13, 2004.
2000 *Mathematics Subject Classification.* 46L05.
Key words and phrases. C*-algebra, amenable trace, representation theory.
This research was supported by MSRI and NSF Postdoctoral Fellowships.

CHAPTER 1

Introduction

One of von Neumann's motivations for initiating the study of operator algebras was to provide an abstract framework for unitary representation theory of locally compact groups. Hence it is no surprise that representation theory of C*-algebras attracted the attention of many experts over the years. However, Glimm's deep work, essentially closing the book on global (irreducible) representation theory, combined with the emergence of exciting new fields such as Connes' noncommutative geometry, Jones' theory of subfactors, Elliott's classification program and, most recently, Voiculescu's theory of free probability, have kept representation theory out of the limelight for the last couple decades.

In this paper we revisit representation theory with the goal of convincing the reader that (a) *finite* representation theory is an important subject where much work is still needed and (b) future advances may well provide the key to unlocking some important open problems. Indeed, in addition to developing some basic general theory, these notes contain the solutions to several problems, give streamlined proofs of (generalizations of) some known results and give new insight into other problems. In contrast to irreducible representations or factor representations of type III (e.g. the celebrated work of Powers) it seems that finite representations have not received the attention they deserve. We hope to lay the groundwork for future study in this direction.

By finite representation theory we mean GNS representations arising from tracial states. It turns out that some traces are better than others and the 'good' ones are precisely those which can be interpreted as invariant means.

DEFINITION 1.1. Let $A \subset B(H)$ be a concretely represented, unital C*-algebra. A state τ on A is called an *amenable trace* if there exists a state ϕ on $B(H)$ such that (1) $\phi|_A = \tau$ and (2) $\phi(uTu^*) = \phi(T)$ for every unitary $u \in A$ and $T \in B(H)$.

Amenable traces really are traces (i.e. $\tau(ab) = \tau(ba)$ for all $a, b \in A$). In more classical language, τ is the restriction of a state which contains A in its centralizer and it is well known that this procedure yields a tracial state. In case one hasn't seen this argument, first note that for every $a \in A$ and *unitary* $u \in A$ we have

$$\tau(au) = \tau(u(au)u^*) = \tau(ua).$$

Since every element $b \in A$ is a linear combination of unitaries, it follows that $\tau(ab) = \tau(ba)$ for all $a, b \in A$.

Next we observe that these traces are analogous to invariant means on groups. Recall that amenable groups are, by definition, those which have an invariant mean – i.e. there exists a state on $L^\infty(G)$ which is invariant under the left translation action of G on $L^\infty(G)$. If $A \subset B(H)$ is a C*-algebra then there is a natural action of the unitary group of A on $B(H)$ given by $T \mapsto uTu^*$, where $T \in B(H)$ is arbitrary and $u \in A$ is unitary. Hence condition (2) in the definition above is precisely the statement that *ϕ is a state on $B(H)$ which is invariant under the action of the unitary group of A*. In other words, τ is an amenable trace if and only if it is the restriction of an invariant mean on $B(H)$.[1]

This notion is well known (under different names) and has already shown its importance (cf. [**23**], [**24**], [**45**], [**47**], [**8**], [**67**], [**6**]). Indeed, in Connes' remarkable paper [**23**] he showed (using the terminology 'hypertrace') that a II_1-factor is isomorphic to the hyperfinite II_1-factor if and only if its unique trace is amenable. Continuing the philosophy that this notion should correspond to some kind of amenability, Bekka defined a unitary representation of a locally compact group to be amenable if the C^*-algebra generated by the image of the representation has an amenable trace (cf. [**8**]). The most important result concerning the structure of amenable traces, however, is due to Kirchberg who (building on work of Connes and using the terminology 'liftable') showed that amenable traces are precisely those which enjoy a natural finite dimensional approximation property (cf. [**47**]). This is reminiscent of the approximation properties which appear in the theories of nuclear, exact or quasidiagonal C^*-algebras and, just as for operator algebras, it proves very useful.

In addition to the study of amenable traces and their corresponding GNS representations, we will also investigate II_1-factor representations of Popa algebras. These algebras are defined via an internal finite dimensional approximation property (see Definition 1.2) which is the C^*-analogue of an approximation property which characterizes the hyperfinite II_1-factor R. In fact, the definition of a Popa algebra appears to be so close to the characterization of R that it was believed for some time that there could only be one II_1-factor which arose from a representation of a Popa algebra (namely R). We will see that this is not the case as the finite representation theory of these algebras is quite rich (containing all McDuff factors, for example).

While we believe that both amenable traces and factor representations of Popa algebras are topics of independent interest, perhaps the most surprising part of this work is that combining these two aspects of representation theory leads to a variety of new results which provide a common thread between several important problems in operator algebras. Our results are most strongly connected to the classification program and questions around free probability but there are also relations with geometric group theory (see Proposition 4.1.13 and questions (8) and (9) in Section 6.7), theoretical numerical analysis and operator theory (see Section 6.4), K-homology (see Section 6.5) and the C^*-algebraic structure of the hyperfinite II_1-factor (see Section 6.6).

Since we deal with a broad range of topics, it may be worthwhile to give a detailed overview before proving any results.

1.1. Approximating Traces on C^*-algebras

As mentioned above, amenable traces are precisely those which arise from a natural finite dimensional approximation property. Strengthening this property in various ways leads to natural definitions of subspaces of the space of amenable traces (see Sections 3.2, 3.3 and 3.4). Studying these stronger approximation properties

---

[1] At this point one may wonder if this notion depends on the particular choice of representation $A \subset B(H)$. It doesn't, as we will observe later.

is roughly the analogue of passing from the class of nuclear C*-algebras to a subclass such as nuclear, quasidiagonal C*-algebras or homogeneous C*-algebras. One subspace of the amenable traces precisely characterizes those GNS representations which have hyperfinite von Neumann algebras after closing in a weak topology (see Theorem 3.2.2). The relation between the other subspaces and representation theory is not yet clear. However, we will see that these other subspaces play a critical role in Elliott's classification program (see Section 6.1).

As mentioned above, basic facts about amenable traces have a number of consequences. Sections 6.4 and 6.5 work out some applications to the finite section method from numerical analysis and some basic K-homological questions, respectively. We also observe that these traces provide natural obstructions to the existence of (unital) *-homomorphisms between certain classes of operator algebras (cf. Corollary 4.2.3).

1.2. II_1-Factor Representations of Popa Algebras

Another goal of these notes is to study II_1-factor representations of Popa algebras.

DEFINITION 1.2. A simple, separable, unital C*-algebra, A, is called a *Popa algebra* if for every finite subset $\mathfrak{F} \subset A$ and $\varepsilon > 0$ there exists a nonzero finite dimensional C*-subalgebra $B \subset A$ with unit e such that $\|ex - xe\| \leq \varepsilon$ for all $x \in \mathfrak{F}$ and $e\mathfrak{F}e \subset^\varepsilon B$ (i.e. for each $x \in \mathfrak{F}$ there exists $b \in B$ such that $\|exe - b\| \leq \varepsilon$).

Popa algebras are always quasidiagonal. Using his local quantization technique in the C*-algebra setting Popa nearly provides a converse in [**67**]: Every simple, unital, quasidiagonal C*-algebra with 'sufficiently many projections' (e.g. real rank zero) is a Popa algebra. Thus the class of Popa algebras is much larger than one might first guess.

For some time there was speculation that quasidiagonality may be closely related to nuclearity. For example, in [**67**, pg. 157] Popa asked whether every Popa algebra with *unique* trace is necessarily nuclear. (Counterexamples were first constructed by Dadarlat in [**26**].) More generally, he asked in [**67**, Remark 3.4.2] whether the hyperfinite II_1-factor R was the *only* II_1-factor which could arise from a GNS representation of a Popa algebra. Support for a positive answer was provided by the following observation of Popa (cf. [**65**], [**66**]):

THEOREM. *Let M be a separable II_1-factor. Then $M \cong R$ if and only if for every finite subset $\mathfrak{F} \subset M$ and $\varepsilon > 0$ there exists a nonzero finite dimensional C*-subalgebra $B \subset M$ with unit e such that $\|ex - xe\|_2 \leq \varepsilon \|e\|_2$ for all $x \in \mathfrak{F}$ and for each $x \in \mathfrak{F}$ there exists $b \in B$ such that $\|exe - b\|_2 \leq \varepsilon \|e\|_2$, where $\|\cdot\|_2$ is the 2-norm on M coming from the unique trace.*

Note that the definition of a Popa algebra is the C*-analogue of the approximation property above which characterizes R. Moreover, the 2-norm version appears, at first glance, to allow one to check the approximation property on a weakly dense subalgebra. In other words, it was quite natural to expect that if a II_1-factor M contained a weakly dense Popa algebra then it would have to be isomorphic to R. Indeed, suppose $A \subset M$ is a weakly dense Popa algebra, $\mathfrak{F} \subset A$, a finite set in the unit ball of A, and $\varepsilon > 0$ are given. By definition we can find a nonzero finite dimensional C*-subalgebra $B \subset A$ with unit e such that $\|ex - xe\| \leq \varepsilon$ for all $x \in \mathfrak{F}$

and $e\mathfrak{F}e \subset^\varepsilon B$. From the general inequality $\|ab\|_2 \leq \|a\|\|b\|_2$ it follows that
$$\|ex - xe\|_2 \leq \|ex - exe\|_2 + \|exe - xe\|_2 = \|e(ex - xe)\|_2 + \|(ex - xe)e\|_2 \leq 2\varepsilon\|e\|_2$$
for all $x \in \mathfrak{F}$ and for each $x \in \mathfrak{F}$ there exists $b \in B$ such that
$$\|exe - b\|_2 = \|e(exe - b)\|_2 \leq \varepsilon\|e\|_2.$$
Hence M has the desired approximation on a weakly dense subalgebra and one "should" be able to pass from A to all of M via some additional argument (which would imply $M \cong R$). Since Popa algebras are simple all their GNS representations are faithful and hence it was expected that R would be the only II_1-factor which could arise from a GNS representation.

It turns out, however, that this is not the case (i.e. it is impossible to use an approximation argument to pass from A to all of M): For example, we will construct a Popa algebra A with the property that for every (separable) II_1-factor M there exists a tracial state τ on A such that $\pi_\tau(A)'' \cong M\bar{\otimes}R$ (see Theorem 5.3.3). Popa has also asked (private communication) if a Popa algebra with unique trace must necessarily yield the hyperfinite II_1-factor. We will see that this question also has a negative answer: There exists a Popa algebra with Dixmier property (hence unique trace) such that the GNS construction yields a non-hyperfinite II_1-factor (see Theorem 6.2.7). On the other hand we will show that if A is a locally reflexive (e.g. exact) Popa algebra with unique trace τ then $\pi_\tau(A)'' \cong R$ thus giving a positive answer to Popa's question in this case (see Theorem 5.3.2).

1.3. Applications

We already mentioned that amenable traces have applications to certain questions in theoretical numerical analysis, K-homology and are related to some natural questions concerning the C*-algebraic structure of the hyperfinite II_1-factor. We now describe, in more detail, connections with some important open problems in other areas of operator algebra theory.

Elliott's Classification Program

There are several new results which those in the classification program may find of interest. For example, we will observe that certain questions around tracial approximation properties provide both necessary conditions and sufficient conditions for various cases of Elliott's conjecture to hold (cf. Propositions 6.1.21 and 6.1.23). As motivation for the other results, we briefly describe the state of affairs (as we see it) in the real rank zero case of the classification program.

That all approximately finite dimensional (AFD) II_1-factors are isomorphic was known to Murray and von Neumann. In [23] Connes proved that all injective II_1-factors are AFD and hence fall under the Murray-von Neumann classification theorem. Elliott's classification program, in our opinion, is in a state similar to the II_1-factor case prior to Connes' work – the classification theorems (for very large classes of real rank zero algebras) exist and we should try to decide whether or not the various hypotheses are always satisfied. Of course, more general classification theorems would always be welcome, but we already have very good results which we should also be trying to exploit.

For example, Huaxin Lin has succeeded in classifying his so-called tracially AF algebras. Roughly speaking, the only difference between a Popa algebra and a (simple) tracially AF algebra is that the finite dimensional algebra B from Definition 1.2 is required (for tracially AF algebras) to be 'large in trace' (see [51] for the precise definition and [**50**] for the classification theorem). Lin's classification result is so exciting because Popa proved in [**67**] that every simple, unital, quasidiagonal C*-algebra of real rank zero is a Popa algebra and hence, 'almost' tracially AF. (Note that Popa requires no nuclearity or K-theoretic hypotheses!) This is the great advantage of tracially AF algebras as opposed to the AH algebras considered in [**30**] as there are no general hypotheses (yet) which allow one to deduce an AH-type structure.

Thus Lin's theorem, combined with Popa's work, may (someday) nearly complete the simple, quasidiagonal, real rank zero case of Elliott's conjecture – there are rather mild K-theoretic restrictions. The K-theory of a tracially AF algebra must be weakly unperforated and satisfy the Riesz interpolation property. However, every such invariant arises from a tracially AF algebra and hence this case of Elliott's conjecture is *equivalent* to the following question (modulo a UCT assumption):

Is every simple, unital, nuclear, quasidiagonal C-algebra with real rank zero, weakly unperforated K-theory and having the Riesz decomposition property necessarily tracially AF?*

Thus the real question is what other general hypotheses (in addition to quasidiagonality and real rank zero) are needed to ensure that the finite dimensional algebras, which exist by Popa's work, can be taken large in trace? Are assumptions on K-theory enough? Some experts have felt that this question would have little to do with nuclearity, primarily because Popa's work requires no such hypotheses. (See, for example, [**51**, page 694] where it was "tempting to conjecture that every quasidiagonal, simple C*-algebra of real rank zero, stable rank one and with weakly unperforated K_0 is tracially AF.") We will see, however, that K-theoretic assumptions are not enough and, moreover, nuclearity must play a role (if the question above is to be resolved affirmatively) as there exists an *exact* Popa algebra with very nice K-theory which is not tracially AF (cf. Corollary 6.2.5). The one drawback of this example is that it does not have a unique tracial state. Popa has asked (private communication) if a unique trace would be enough to ensure that Popa algebras are tracially AF. We will see that this is also not enough (see Theorem 6.2.7). It is interesting to note, however, that the case of an exact Popa algebra with unique trace remains open.

On the other hand, inspired by recent work of Huaxin Lin [**53**], we will show that approximation properties of traces provide an abstract hypothesis which does ensure that Popa algebras are tracially AF. That is, under certain technical assumptions, we will show that (not necessarily nuclear) tracially AF algebras are easily *characterized* by tracial approximation properties (see Proposition 4.5.5). Moreover, in the presence of nuclearity and a unique trace it may turn out that this approximation property is always satisfied (see the discussion in Section 6.1). It is the case that type I C*-algebras *always* have this nice tracial approximation property and hence we are able to classify many C*-algebras which are built up out of type I algebras (for example, thanks to the deep work of Q. Lin and Phillips, all crossed products of compact manifolds by diffeomorphisms which have real rank zero and unique trace). Our classification results are similar to recent work of Huaxin Lin, however the proofs are significantly shorter and the present approach strikes us as technically and conceptually simpler.

Free Probability

There are two new results arising from this work which are relevant to free probability. One is related to Connes' embedding problem and the other is related to the semicontinuity and invariance questions for Voiculescu's free entropy dimension.

Regarding Connes' embedding problem (i.e. the question of whether or not microstates always exist) we obtain the following result (see Theorem 6.3.1): A II_1-factor M embeds into the ultraproduct of the hyperfinite II_1-factor if and only if there exists a weakly dense C*-subalgebra $A \subset M$ and a u.c.p. map $\Phi : B(L^2(M)) \to M$ such that $\Phi(a) = a$ for all $a \in A$ (i.e. M has Lance's WEP relative to a weakly dense subalgebra). We also show that this is equivalent to the following: for every finite subset $\mathfrak{F} \subset M$ and $\varepsilon > 0$ there exists an operator system $X \subset M$ such that $\mathfrak{F}$ is ε-contained in X (in 2-norm) and X is completely order isomorphic to the hyperfinite II_1-factor (i.e. M is quite literally built out of the hyperfinite II_1-factor). Hence Connes' embedding problem predicts that the hyperfinite II_1-factor is *the* basic building block for all II_1-factors. Our own feeling is that the world of II_1-factors is too exotic to expect that everything is built up out of the nicest possible II_1-factor, but we have not yet been able to construct a counterexample. On the other hand, this result also shows that many well known II_1-factors (e.g. free group factors or property T factors coming from residually finite groups) have a *dense* internal structure reminiscent of the hyperfinite II_1-factor (but, of course, are not themselves hyperfinite).[2]

As previously mentioned, we will show that every McDuff factor contains a weakly dense Popa algebra (Theorem 5.3.3). Based on the proof of this result, in joint work with Ken Dykema, we have been able to show that all the (interpolated) free group factors $L(\mathbb{F}_n)$ ($1 < n < \infty$) contain finitely generated, weakly dense Popa algebras (see [**19**]). The point is that Popa algebras are fundamentally different, at least from a C*-algebraic point of view, than the canonical generators of free group factors. For example, Popa algebras *never* contain two Haar unitaries which are free with respect to some tracial state.[3] In other words, the techniques of this paper lead to new constructions of generators of some well known II_1-factors. This is relevant to the invariance question for free entropy dimension as all new generators give candidates for computation. We make no attempt to compute free entropy dimension in this paper – our only point is that the results presented here give new tools to construct "exotic" generators of well known II_1-factors and we hope our work provides the foundation for more examples in the future.

ACKNOWLEDGEMENT. This work began during the year-long program in operator algebras at MSRI, 2000-2001. I spoke to nearly everyone I encountered during that year about various aspects of this work, as well as most everyone I have encountered since, and it would be impossible to recall all of the people who contributed remarks and ideas. Instead I express my sincerest thanks to MSRI, the organizers and participants and everyone else over the last three years who patiently endured conversations on this topic. However, I must specifically thank Marius Dadarlat

---

[2]Though it is standard, it is a bit misleading to use the terminology "hyperfinite" in this paragraph as "injective" would be more appropriate. Of course, Connes' uniqueness theorem asserts that these notions coincide for II_1-factors, but the isomorphism identifying R with a subspace $X \subset M$ is not normal and hence the finite dimensional subalgebras which are weakly dense in R will map over to some small portion of X. However, injectivity is preserved so that X is still an injective operator system.

[3]This is because such unitaries would generated a copy of $C_r^*(\mathbb{F}_2)$, but Popa algebras are quasidiagonal and hence can't contain a copy of the non-quasidiagonal C*-algebra $C_r^*(\mathbb{F}_2)$.

and Dimitri Shlyakhtenko for a (seemingly infinite) number of helpful discussions. Finally I thank the University of Tokyo, Yasu Kawahigashi and Taka Ozawa for their hospitality as a significant part of the (re)writing of these notes took place during a one year visiting position at the University of Tokyo.

CHAPTER 2

Notation, definitions and useful facts

The purpose of this chapter is to set our notation and list a number of facts which will be used throughout. Our notation should be standard and the facts we collect here are all well known, very simple or minor variations of the statements most commonly seen in the literature. In particular, there should be no harm in jumping to the next chapter and referring back as necessary.

Unless otherwise noted or obviously false, all C*-algebras are assumed to be unital and separable. Similarly, all von Neumann algebras will be assumed to have separable preduals (with the exception of R^ω, which is well known to be non-separable).

For a Hilbert space H, we will let $B(H)$ and $\mathcal{K}(H)$ denote the bounded and, respectively, compact operators on H. The canonical (unbounded, densely defined) trace on $B(H)$ will be denoted by Tr while $\|\cdot\|$ will be the operator norm on $B(H)$, $\|\cdot\|_{HS}$ and $\|\cdot\|_1$ will be the Hilbert-Schmidt and L^1-norms, respectively[1], and $\langle\cdot,\cdot\rangle_{HS}$ will be the inner product on the Hilbert space of Hilbert-Schmidt operators.[2]

When A is a C*-algebra with state η we will denote the associated GNS Hilbert space, representation and von Neumann algebra by $L^2(A,\eta)$, $\pi_\eta : A \to B(L^2(A,\eta))$ and $\pi_\eta(A)''$, respectively. Given $a \in A$, $\hat{a} \in L^2(A,\eta)$ will be the canonical image of a in the GNS Hilbert space.

The symbols $\odot$, $\otimes$ and $\bar{\otimes}$ will denote the algebraic, minimal and W*-tensor products, respectively.

If A is a C*-algebra we will let A^{op} denote the opposite algebra (i.e. $A^{op} = A$ as involutive normed linear spaces, but multiplication in A^{op} is defined by $a \circ b = ba$; the latter multiplication being the given multiplication in A). A^{**} will denote the enveloping von Neumann algebra of A (i.e. the Banach space double dual of A). Contrary to our standing assumption that von Neumann algebras should have separable preduals, A^{**} usually has a non-separable predual.

If $\tau \in \mathrm{T}(A)$ is a tracial state then there is a canonical antilinear isometry $J : L^2(A,\tau) \to L^2(A,\tau)$ defined by $J(\hat{a}) = \widehat{a^*}$. One defines a 'right regular representation' (i.e. a *-homomorphism $\pi_\tau^{op} : A^{op} \to B(L^2(A,\tau))$) by $\pi_\tau^{op}(a) = J\pi_\tau(a^*)J$. Since $J\pi_\tau(A)J \subset \pi_\tau(A)'$ one then gets an algebraic homomorphism $\pi_\tau \odot \pi_\tau^{op} : A \odot A^{op} \to B(L^2(A,\tau))$ defined on elementary tensors by $\pi_\tau \odot \pi_\tau^{op}(a \otimes b) = \pi_\tau(a)\pi_\tau^{op}(b)$. It is an important fact, essentially due to Murray and von Neumann, that $J\pi_\tau(A)''J = \pi_\tau(A)'$ and (hence) $\pi_\tau(A)'' = J\pi_\tau^{op}(A)'J$.

Completely positive maps (cf. [60]) will play an important role in these notes. We will use the abbreviations c.p. and u.c.p. for 'completely positive' and 'unital

---

[1] i.e. $\|T\|_{HS}^2 = \mathrm{Tr}(T^*T)$ and $\|T\|_1 = \mathrm{Tr}(|T|)$.
[2] i.e. $\langle S,T\rangle_{HS} = \mathrm{Tr}(T^*S)$.

completely positive', respectively. We will often use two fundamental results concerning such maps: Stinespring's dilation theorem and Arveson's extension theorem (cf. [**60**]).

Multiplicative domains of u.c.p. maps will appear several times. For a u.c.p. map $\phi : A \to B$ we will let A_ϕ denote the multiplicative domain (cf. [**60**]). By definition,
$$A_\phi = \{a \in A : \phi(a^*a) = \phi(a^*)\phi(a) \text{ and} \phi(aa^*) = \phi(a)\phi(a^*)\}$$
and it is not too hard to show that this set can also be described as the set of elements which commute with the Stinespring projection (in the Stinespring representation). The key fact for us is that *u.c.p. maps are bimodule maps over their multiplicative domains* – i.e. if $a, c \in A_\phi$ and $b \in A$ then $\phi(abc) = \phi(a)\phi(b)\phi(c)$.

The hyperfinite II_1-factor will appear many times and will always be denoted by R. We will use tr_n to denote the unique tracial state on the $n \times n$ matrices. For a von Neumann algebra, M, with faithful, normal, tracial state τ, we will let $\|\cdot\|_{2,\tau}$ be the associated 2-norm (i.e. $\|x\|_{2,\tau} = \tau(x^*x)^{1/2}$). If M has a unique trace (i.e. is a factor) then we will drop the dependence on τ and simply write $\|\cdot\|_2$.

The ultraproduct of the hyperfinite II_1 factor R^ω will also appear several times. That is, given a free ultrafilter $\omega \in \beta\mathbb{N}\setminus\mathbb{N}$ one defines an ideal $I_\omega \subset l^\infty(R) = \{(x_n) \in \Pi_{n \in \mathbb{N}} R : \sup_{n \in \mathbb{N}} \|x_n\| < \infty\}$ by $I_\omega = \{(x_n) \in l^\infty(R) : \lim_{n \to \omega} \|x_n\|_2 = 0\}$. Then the ultraproduct of R with respect to ω is defined to be the (C*-algebraic) quotient: $R^\omega = l^\infty(R)/I_\omega$. R^ω is a II_1 factor with trace $\tau_\omega((x_n) + I_\omega) = \lim_{n \to \omega} \tau_R(x_n)$.

The following simple fact will be useful.

LEMMA 2.1. *Let $A \subset B(H)$ be a C*-algebra and $P \in B(H)$ be a finite rank projection. Then*
$$\|Pa - aP\| = \max\{\|Paa^*P - PaPa^*P\|^{1/2}, \|Pa^*aP - Pa^*PaP\|^{1/2}\},$$
and
$$\frac{\|Pa - aP\|_{HS}}{\|P\|_{HS}} = \left(\frac{\mathrm{Tr}(Paa^*P - PaPa^*P) + \mathrm{Tr}(Pa^*aP - Pa^*PaP)}{\mathrm{Tr}(P)}\right)^{1/2}.$$

PROOF. Use the identity
$$Pa - aP = Pa(1 - P) - (1 - P)aP,$$
the fact that $Pa(1-P)$ and $(1-P)aP$ have orthogonal domains and ranges (hence are orthogonal Hilbert-Schmidt vectors) and compute away. $\square$

We will need the following version of Voiculescu's Theorem.

THEOREM 2.2. *Let $A \subset B(H)$ be in general position (i.e. $A \cap \mathcal{K}(H) = \{0\}$). If $\phi : A \to M_n(\mathbb{C})$ is a u.c.p. map then there exist isometries $V_k : \mathbb{C}^n \to H$ such that $\|\phi(a) - V_k^*aV_k\| \to 0$, for all $a \in A$, as $k \to \infty$. Moreover, letting $P_k = V_kV_k^*$, we have*
$$\lim_{k \to \infty} \|P_ka - aP_k\| = \max\{\|\phi(aa^*) - \phi(a)\phi(a^*)\|^{\frac{1}{2}}, \|\phi(a^*a) - \phi(a^*)\phi(a)\|^{\frac{1}{2}}\}$$
and
$$\lim_{k \to \infty} \frac{\|P_ka - aP_k\|_{HS}}{\|P_k\|_{HS}} = \left(\mathrm{tr}_n(\phi(aa^*) - \phi(a)\phi(a^*)) + \mathrm{tr}_n(\phi(a^*a) - \phi(a^*)\phi(a))\right)^{\frac{1}{2}}.$$

PROOF. That there exist isometries $V_k : \mathbb{C}^n \to H$ such that $\|\phi(a) - V_k^* a V_k\| \to 0$ is the first step in proving the usual version of Voiculescu's Theorem (cf. [**3**] or [**27**]). The commutator estimates, which are the important part for us, follow from the previous lemma. □

We will also need the following technical version of Voiculescu's Theorem. A proof can be found in [**17**] or [**18**].

PROPOSITION 2.3. *Let $A \subset B(H)$ be in general position and $\Phi : A \to B(K)$ be a u.c.p. map which is a faithful $*$-homomorphism modulo the compacts (i.e. composing with the quotient map to the Calkin algebra yields a faithful $*$-monomorphism $A \hookrightarrow Q(K)$). Then there exists a sequence of unitaries $U_n : K \to H$ such that for every $a \in A$ we have*

$$\limsup \|a - U_n \Phi(a) U_n^*\| \leq 2\max\{\|\Phi(aa^*) - \Phi(a)\Phi(a^*)\|^{\frac{1}{2}}, \|\Phi(a^*a) - \Phi(a^*)\Phi(a)\|^{\frac{1}{2}}\}.$$

We now list four simple facts which will also be useful. The first two are well known and the second two are just a bit of trickery. We begin with a simple adaptation of the fact that if M is a von Neumann algebra with faithful, normal tracial state τ and $1_M \in N \subset M$ is a sub-von Neumann algebra then there always exists a τ-preserving (hence faithful) conditional expectation $M \to N$ (cf. [**44**, Exercise 8.7.28]).

LEMMA 2.4. *Let A be a C^*-algebra, $\tau \in \mathrm{T}(A)$ be a tracial state and $1_A \in B \subset A$ be a finite dimensional subalgebra. Then there exists a conditional expectation $\Phi_B : A \to B$ such that $\tau \circ \Phi_B = \tau$.*

PROOF. Assume first that $\tau|_B$ is faithful. Write $B \cong M_{n(1)}(\mathbb{C}) \oplus \cdots \oplus M_{n(k)}(\mathbb{C})$, and let $\{e_{i,j}^{(1)}\}_{1 \leq i,j \leq n(1)} \cup \ldots \cup \{e_{i,j}^{(k)}\}_{1 \leq i,j \leq n(k)}$ be a system of matrix units for B. Then the desired conditional expectation is given by

$$\Phi_B(x) = \sum_{s=1}^{k} \sum_{i,j=1}^{n(s)} \frac{\tau(x e_{i,j}^{(s)})}{\tau(e_{i,i}^{(s)})} e_{j,i}^{(s)}.$$

When $\tau|_B$ is not faithful, the formula above no longer makes sense. However, one can decompose B as the direct sum of two finite dimensional algebras, $B_0 \oplus B_f$, where $\tau|_{B_0} = 0$ and $\tau|_{B_f}$ is faithful. Letting e_0 (resp. e_f) be the unit of B_0 (resp. B_f) we get a τ-preserving conditional expectation by mapping each $a \in A$ to $E_{B_0}(e_0 a e_0) + E_{B_f}(e_f a e_f)$, where $E_{B_0} : e_0 A e_0 \to B_0$ is any conditional expectation (which exists by finite dimensionality) and $E_{B_f} : e_f A e_f \to B_f$ is a $\tau|_{e_f A e_f}$ preserving conditional expectation as in the first part of the proof. □

LEMMA 2.5. *If B is a finite dimensional C^*-algebra with tracial state τ then for every $\varepsilon > 0$ there exists $n \in \mathbb{N}$ and a unital $*$-monomorphism $\rho : B \hookrightarrow M_n(\mathbb{C})$ such that $|\tau(x) - \mathrm{tr}_n \circ \rho(x)| < \varepsilon \|x\|$ for every $x \in B$.*

PROOF. If τ is a *rational* convex combination of extreme traces then one can find an honestly trace preserving embedding by inflating the summands of B according to the rational numbers appearing in the convex combination. The general case then follows by approximation. □

We remind the reader of the following theorem of Voiculescu (cf. [**77**, Theorem 1]): A separable, unital C^*-algebra A is quasidiagonal if and only if there exists a sequence of u.c.p. maps $\varphi_n : A \to M_{k(n)}(\mathbb{C})$ which are asymptotically multiplicative (i.e. $\|\varphi_n(ab) - \varphi_n(a)\varphi_n(b)\| \to 0$ for all $a, b \in A$) and asymptotically isometric (i.e. $\|a\| = \lim \|\varphi_n(a)\|$, for all $a \in A$).

LEMMA 2.6. *Assume that A is a quasidiagonal C^*-algebra and $\psi_n : A \to M_{l(n)}(\mathbb{C})$ is an asymptotically multiplicative (but not necessarily asymptotically isometric) sequence of u.c.p. maps. Then there exists a sequence of u.c.p. maps $\Phi_n : A \to M_{t(n)}(\mathbb{C})$ which are asymptotically multiplicative, asymptotically isometric and such that $|\mathrm{tr}_{t(n)}(\Phi_n(a)) - \mathrm{tr}_{l(n)}(\psi_n(a))| \to 0$ for all $a \in A$.*

PROOF. Let $\varphi_n : A \to M_{k(n)}(\mathbb{C})$ be an asymptotically multiplicative, asymptotically isometric sequence of u.c.p. maps. Choose integers $s(n)$ such that $\frac{k(n)}{s(n)} \to 0$; thus $\frac{s(n)l(n)}{s(n)l(n)+k(n)} \to 1$. Then one defines $\Phi_n : A \to M_{s(n)l(n)+k(n)}(\mathbb{C})$ to be the block diagonal map with one summand equal to φ_n and $s(n)$ summands equal to ψ_n. □

Note that the previous lemma can be formulated in terms of ∗-homomorphisms when A is a residually finite dimensional C^*-algebra.

Though we will try to keep everything unital, non-unital maps are sometimes unavoidable. The next lemma keeps everything running smoothly.

LEMMA 2.7. *Let $\phi_n : A \to M_{k(n)}(\mathbb{C})$ be contractive c.p. (c.c.p.) maps (though we still assume A is unital).*

(1) *If $\|\phi_n(ab) - \phi_n(a)\phi_n(b)\|_2 \to 0$ for all $a, b \in A$ and $\mathrm{tr}_{k(n)}(\phi_n(1_A)) \to 1$ then there exist u.c.p. maps $\psi_n : A \to M_{k(n)}(\mathbb{C})$ which are also asymptotically multiplicative with respect to 2-norms and such that $|\mathrm{tr}_{k(n)}(\psi_n(a)) - \mathrm{tr}_{k(n)}(\phi_n(a))| \to 0$ as $n \to \infty$, with convergence being uniform on the unit ball of A.*

(2) *If $\|\phi_n(ab) - \phi_n(a)\phi_n(b)\| \to 0$ for all $a, b \in A$ and $\mathrm{tr}_{k(n)}(\phi_n(1_A)) \to 1$ then there exist integers $l(n) \leq k(n)$ and asymptotically multiplicative u.c.p. maps $\psi_n : A \to M_{l(n)}(\mathbb{C})$ such that $|\mathrm{tr}_{l(n)}(\psi_n(a)) - \mathrm{tr}_{k(n)}(\phi_n(a))| \to 0$ as $n \to \infty$, with convergence being uniform on the unit ball of A.*

PROOF. For the proof of (1) we will need an observation of Choi-Effros (cf. [**22**, Lemma 2.2]) which says that we can find u.c.p. maps $\psi_n : A \to M_{k(n)}(\mathbb{C})$ such that $\phi_n(a) = c_n \psi_n(a) c_n$ for all n and $a \in A$, where $c_n = \phi_n(1_A)^{1/2}$. Since $\mathrm{tr}_{k(n)}(\phi_n(1_A)) \to 1$ it follows that $\|c_n - 1_{M_{k(n)}}\|_2 \to 0$. Also we note the following general inequality for all $x \in M_{k(n)}(\mathbb{C})$:

$$\|x - c_n x c_n\|_2 \leq 2\|x\| \|c_n - 1_{M_{k(n)}}\|_2.$$

Hence we find that

$$\begin{aligned}
\|\psi_n(ab) - \psi_n(a)\psi_n(b)\|_2 &\leq \|\psi_n(ab) - c_n \psi_n(ab) c_n\|_2 + \|\phi_n(ab) - \phi_n(a)\phi_n(b)\|_2 \\
&\quad + \|\phi_n(a)\phi_n(b) - \psi_n(a)\psi_n(b)\|_2 \\
&\leq 2\|ab\| \|c_n - 1_{M_{k(n)}}\|_2 + \|\phi_n(ab) - \phi_n(a)\phi_n(b)\|_2 \\
&\quad + 4\|a\|\|b\| \|c_n - 1_{M_{k(n)}}\|_2 \\
&\leq 6\|a\|\|b\| \|c_n - 1_{M_{k(n)}}\|_2 + \|\phi_n(ab) - \phi_n(a)\phi_n(b)\|_2.
\end{aligned}$$

This shows asymptotic multiplicativity of the ψ_n's.

The tracial approximation part is a simple application of the Cauchy-Schwartz inequality: since $|\text{tr}_{k(n)}(\psi_n(a)) - \text{tr}_{k(n)}(c_n\psi_n(a)c_n)| = |\text{tr}_{k(n)}(\psi_n(a)(1_{M_{k(n)}} - c_n^2))|$ we see that

$$|\text{tr}_{k(n)}(\psi_n(a)) - \text{tr}_{k(n)}(c_n\psi_n(a)c_n)| \leq \|a\|\|c_n^2 - 1_{M_{k(n)}}\|_2.$$

The proof of part (2) uses similar estimates as those above but starts off a little different. Hence we only describe how to get the ψ_n's and leave the remaining details to the reader. In the case that $\|\phi_n(ab) - \phi_n(a)\phi_n(b)\| \to 0$ we have that the $\phi_n(1_A)$'s are getting closer and closer to projections (by functional calculus), say P_n. One checks that the c.c.p. maps $A \to M_{l(n)}(\mathbb{C})$ defined by $a \mapsto P_n\phi_n(a)P_n$ are also asymptotically multiplicative – $\|P_n\phi_n(a) - \phi_n(a)P_n\| \to 0$, for all $a \in A$, since $\phi_n(1_A)$ is asymptotically central. Unfortunately they still aren't unital, but $\phi_n(1)$ is invertible in $P_n M_{k(n)}(\mathbb{C})P_n$, since it is very close in norm to the identity, and hence we get u.c.p. maps by defining

$$\psi_n(a) = (P_n\phi_n(1))^{-\frac{1}{2}}(P_n\phi_n(a)P_n)(P_n\phi_n(1))^{-\frac{1}{2}}.$$

That these maps ψ_n are still asymptotically multiplicative in norm is similar to the 2-norm case above as is the tracial approximation property. □

Finally, we will need the following consequence of the Hahn-Banach theorem. This result is a good exercise (for those not already familiar with it).

LEMMA 2.8. *Let $S(A)$ denote the state space of a C^*-algebra A. Assume that $\mathcal{S} \subset S(A)$ is a set of states with the property that for each self-adjoint $a \in A$ we have*

$$\|a\| = \sup_{\phi \in \mathcal{S}}\{|\phi(a)|\}.$$

Then the (weak-) closed, convex hull of $\mathcal{S}$ is equal to $S(A)$.*

CHAPTER 3

Amenable traces and stronger approximation properties

We will now develop some of the basic theory of amenable traces. Our first goal is to show that this notion has numerous equivalent formulations, just as amenable groups admit numerous characterizations. These fundamental results, which are essentially due to Connes in the unique trace case and Kirchberg in general, are non-trivial and have already appeared in the literature (cf. [**23**], [**47**]). However, due to its importance, we feel that a detailed presentation of Ozawa's (substantially simplified but still quite technical) proof is worthwhile (cf. [**58**]).

As previously mentioned, one of the characterizations of amenable traces is a natural finite dimensional approximation property. In later sections we will define several subsets of the amenable traces by simply strengthening this approximation property in various ways. At this point our goal is just to define these subsets and make a few general observations. We will discuss examples and applications in the later.

We remind the reader that, unless obviously false, all C*-algebras (resp. von Neumann algebras) are assumed to be separable and unital (resp. have separable preduals). We also remind the reader that all notation not defined below should have been explained in the previous chapter.

3.1. Characterizations of amenable traces

We first recall the definition of an amenable trace.

DEFINITION 3.1.1. Let $A \subset B(H)$ be a concretely represented, unital C*-algebra. A state τ on A is called an *amenable trace* if there exists a state ϕ on $B(H)$ such that (1) $\phi|_A = \tau$ and (2) $\phi(uTu^*) = \phi(T)$ for every unitary $u \in A$ and $T \in B(H)$. We will denote the set of amenable traces on A by AT(A).

Though it appears that this set of traces may depend on the choice of representation $A \subset B(H)$ it turns out that this is not the case.

PROPOSITION 3.1.2. *Let $A \subset B(H)$ be a C*-algebra, $\tau \in$ AT(A) and $\pi : A \to B(K)$ be any other faithful representation of A. Then $\tau \circ \pi^{-1}$ is an amenable trace on $\pi(A)$. In other words, the set AT(A) does not depend on the choice of faithful representation $A \subset B(H)$ and hence being an amenable trace is a natural abstract property of a tracial state.*

PROOF. By Arveson's extension theorem we can find a u.c.p. map $\Psi : B(K) \to B(H)$ such that $\Psi(\pi(a)) = a$ for all $a \in A$. Note that $\pi(A)$ belongs to the multiplicative domain of Ψ since $\Psi|_{\pi(A)}$ is a *-homomorphism (namely, π^{-1}). Define a state ψ on $B(K)$ by $\psi = \phi \circ \Psi$. Now for arbitrary $T \in B(K)$ and unitary $u \in A$

we have

$$\psi(\pi(u)T\pi(u^*)) = \phi(\Psi(\pi(u))\Psi(T)\Psi(\pi(u^*))) = \phi(u\Psi(T)u^*) = \phi(\Psi(T)) = \psi(T).$$

Hence $\tau \circ \pi^{-1}$ is an amenable trace on $\pi(A)$. □

We now begin the long and technical, though essentially elementary, journey to the various characterizations of amenable traces. This will require numerous calculations[1] as well as the *Powers-Størmer* inequality.

PROPOSITION 3.1.3 (Powers-Størmer). *If $a, b \in B(H)$ are positive trace class operators then $\|a-b\|_{HS}^2 \leq \|a^2 - b^2\|_1$. In particular, if $u \in B(H)$ is a unitary and $h \geq 0$ has finite rank then $\|uh^{1/2} - h^{1/2}u\|_{HS} = \|uh^{1/2}u^* - h^{1/2}\|_{HS} \leq \|uhu^* - h\|_1^{1/2}$.*

PROOF. We could refer to the original paper, of course, but the proof is short and elementary so we include it.

Let $\{v_k\}$ be an orthonormal basis of H consisting of eigenvectors of $a - b$ and let λ_k be the corresponding (real) eigenvalues. Note that since $a + b \geq a - b$ and $a + b \geq -(a - b)$ it follows that

$$\langle (a+b)v, v \rangle \geq |\langle (a-b)v, v \rangle|,$$

for all $v \in H$. Note also that for any self-adjoint x we have the inequality $|\langle xv, v \rangle| \leq \langle |x|v, v \rangle$.

Now we compute

$$\begin{aligned}
\|a - b\|_{HS}^2 &= \text{Tr}(|a-b|^2) \\
&= \sum_k \langle |a-b|^2 v_k, v_k \rangle \\
&= \sum_k |\lambda_k|^2 \\
&\leq \sum_k |\lambda_k| \langle (a+b)v_k, v_k \rangle \\
&= \sum_k |\langle (a+b)(\tfrac{1}{2}\lambda_k v_k), v_k \rangle + \langle (a+b)v_k, \tfrac{1}{2}\lambda_k v_k \rangle| \\
&= \sum_k |\langle \tfrac{1}{2}((a+b)(a-b) + (a-b)(a+b))v_k, v_k \rangle| \\
&= \sum_k |\langle (a^2 - b^2)v_k, v_k \rangle| \\
&\leq \sum_k \langle |a^2 - b^2| v_k, v_k \rangle \\
&= \|a^2 - b^2\|_1.
\end{aligned}$$

□

Our next lemma is the key technical ingredient. The proof given below is due to Ozawa [**58**, Theorem 6.1] – with a few more details thrown in.

---

[1]The diligent reader should go find a very large chalkboard before reading further!

LEMMA 3.1.4. *Let $h \in B(H)$ be a positive, finite rank operator with rational eigenvalues and $\mathrm{Tr}(h) = 1$. Then there exists a u.c.p. map $\phi : B(H) \to M_q(\mathbb{C})$ such that $\mathrm{tr}_q(\phi(T)) = \mathrm{Tr}(hT)$ for all $T \in B(H)$ and $|\mathrm{tr}_q(\phi(uu^*) - \phi(u)\phi(u^*))| \leq 2\|uhu^* - h\|_1^{1/2}$ for every unitary operator $u \in B(H)$.*

PROOF. Let $v_1, \ldots, v_k \in H$ be the eigenvectors of h and $\frac{p_1}{q}, \ldots, \frac{p_k}{q}$ the corresponding eigenvalues. Note that $\sum p_j = q$ since $\mathrm{Tr}(h) = 1$. Let $\{w_m\}$ be any orthonormal basis of H and consider the following orthonormal subset of $H \otimes H$:

$$\{v_1 \otimes w_1, \ldots, v_1 \otimes w_{p_1}\} \cup \{v_2 \otimes w_1, \ldots, v_2 \otimes w_{p_2}\} \cup \ldots \cup \{v_k \otimes w_1, \ldots, v_k \otimes w_{p_k}\}.$$

Let $P \in B(H \otimes H)$ be the orthogonal projection onto the span of these vectors.

Our first task is to write down the matrix of $P(T \otimes 1)P$ (in the basis above), for an arbitrary $T \in B(H)$. Though we have no intention of doing this completely we will make a few remarks.

The matrix of $P(T \otimes 1)P$ decomposes into $k \times k$ blocks (which are not square!) as follows

$$P(T \otimes 1)P = \begin{pmatrix} A_{1,1} & A_{1,2} & \cdots & A_{1,k} \\ A_{2,1} & A_{2,2} & \cdots & A_{2,k} \\ \vdots & \vdots & \ddots & \vdots \\ A_{k,1} & A_{k,2} & \cdots & A_{k,k} \end{pmatrix}$$

where the matrix $A_{i,j}$ has p_i rows and p_j columns. The matrices $A_{i,i}$ look like

$$\begin{pmatrix} \langle Tv_i, v_i \rangle & 0 & \cdots & 0 \\ 0 & \langle Tv_i, v_i \rangle & \cdots & 0 \\ \vdots & \vdots & \ddots & \vdots \\ 0 & 0 & \cdots & \langle Tv_i, v_i \rangle \end{pmatrix}.$$

For $i \neq j$ the matrices $A_{i,j}$ look like

$$\begin{pmatrix} \langle Tv_j, v_i \rangle & 0 & \cdots & 0 \\ 0 & \langle Tv_j, v_i \rangle & \cdots & 0 \\ \vdots & \vdots & \ddots & \vdots \end{pmatrix},$$

but are not necessarily square (unless $p_i = p_j$). In particular note that the number of $\langle Tv_j, v_i \rangle$'s appearing is equal to $\min\{p_i, p_j\}$.

Now one computes the matrix of $P(T \otimes 1)P(T^* \otimes 1)P$. Having done so the following facts become obvious.

(1) $\frac{1}{q}\mathrm{Tr}(P(T \otimes 1)P) = \frac{1}{q}\left(\sum_{i=1}^k p_i \langle Tv_i, v_i \rangle\right) = \sum_{i=1}^k \langle Thv_i, v_i \rangle = \mathrm{Tr}(Th)$.
(2) $\frac{1}{q}\mathrm{Tr}(P(T \otimes 1)P(T^* \otimes 1)P) = \frac{1}{q}\sum_{i,j=1}^k |\langle Tv_j, v_i \rangle|^2 \min\{p_i, p_j\}$.

As if that wasn't bad enough, one should now write down the matrices of $h^{1/2}T$, $h^{1/2}T^*$ and $h^{1/2}Th^{1/2}T^*$ (in any orthonormal basis which begins with $\{v_1, \ldots, v_k\}$). Having done so one immediately sees that, letting $T_{i,j} = \langle Tv_j, v_i \rangle$,

$$\mathrm{Tr}(h^{1/2}Th^{1/2}T^*) = \sum_{i,j=1}^k \frac{1}{q}(p_i p_j)^{1/2} |T_{i,j}|^2.$$

Hence, if we define a u.c.p. map $\phi : B(H) \to M_q(\mathbb{C})$ by $\phi(T) = P(T \otimes 1)P$ then $\mathrm{tr}_q(\phi(T)) = \mathrm{Tr}(hT)$ for all $T \in B(H)$. For TeXnical reasons we define $(*) =$

$|\operatorname{Tr}(h^{1/2}Th^{1/2}T^*) - \operatorname{tr}_q(\phi(T)\phi(T^*))|$ and observe the following estimates:

$$\begin{aligned}
(*) &= \sum_{i,j=1}^{k} \frac{1}{q}|T_{i,j}|^2\Big((p_ip_j)^{1/2} - \min\{p_i,p_j\}\Big) \\
&\leq \sum_{i,j=1}^{k} \frac{1}{q}|T_{i,j}|^2 p_i^{1/2}|p_i^{1/2} - p_j^{1/2}| \\
&\leq \Big(\sum_{i,j=1}^{k} \frac{1}{q}|T_{i,j}|^2 p_i\Big)^{1/2} \Big(\sum_{i,j=1}^{k} \frac{1}{q}|T_{i,j}|^2 (p_i^{1/2} - p_j^{1/2})^2\Big)^{1/2} \\
&= \|Th^{1/2}\|_{HS} \|h^{1/2}T - Th^{1/2}\|_{HS}.
\end{aligned}$$

Now if T happens to be a unitary operator then $\|Th^{1/2}\|_{HS} = \|h^{1/2}\|_{HS} = 1$ and $\|h^{1/2}T - Th^{1/2}\|_{HS} = \|Th^{1/2}T^* - h^{1/2}\|_{HS}$ and hence we can apply the Powers-Størmer inequality after the inequalities above to get:

$$|\operatorname{Tr}(h^{1/2}Th^{1/2}T^*) - \operatorname{tr}(\phi(T)\phi(T^*))| \leq \|ThT^* - h\|_1^{1/2}.$$

Finally, the Cauchy-Schwartz inequality applied to the Hilbert-Schmidt operators implies that for every unitary operator $T \in B(H)$,

$$\begin{aligned}
\operatorname{tr}_q(\phi(TT^*) - \phi(T)\phi(T^*)) &\leq |1 - \operatorname{Tr}(h^{1/2}Th^{1/2}T^*)| + \|ThT^* - h\|_1^{1/2} \\
&= |\operatorname{Tr}(ThT^*) - \operatorname{Tr}(h^{1/2}Th^{1/2}T^*)| + \|ThT^* - h\|_1^{1/2} \\
&= |\operatorname{Tr}((Th^{1/2} - h^{1/2}T)h^{1/2}T^*)| + \|ThT^* - h\|_1^{1/2} \\
&\leq \|h^{1/2}T^*\|_{HS} \|Th^{1/2} - h^{1/2}T\|_{HS} + \|ThT^* - h\|_1^{1/2} \\
&\leq 2\|ThT^* - h\|_1^{1/2}.
\end{aligned}$$

□

LEMMA 3.1.5. *If $\phi: A \to M_n(\mathbb{C})$ is a u.c.p. map then for all $a, b \in A$ we have*

$$\|\phi(ab) - \phi(a)\phi(b)\|_2 \leq \|b\|\Big(\operatorname{tr}_n\big(\phi(aa^*) - \phi(a)\phi(a^*)\big) + \operatorname{tr}_n\big(\phi(a^*a) - \phi(a^*)\phi(a)\big)\Big)^{1/2}.$$

PROOF. Thanks to Stinespring's dilation theorem we may assume that $A \subset B(H)$ and $\phi(a) = PaP$, for all $a \in A$, where $P \in B(H)$ is a finite rank projection. (If Stinespring's representation is not faithful just dilate it further.) Note that tr_n on $M_n(\mathbb{C}) = PB(H)P$ is then identified with the linear functional $\frac{\operatorname{Tr}(\cdot)}{\operatorname{Tr}(P)}$ and hence the 2-norm $\|\cdot\|_2$ gets identified with

$$\frac{\|\cdot\|_{HS}}{\|P\|_{HS}}.$$

Hence, if $a, b \in A$ are given we can apply Lemma 2.1 to $\|\phi(ab) - \phi(a)\phi(b)\|_2 = (*)$ (again, for TeXnical reasons) and deduce that

$$
\begin{aligned}
(*) &= \frac{\|PabP - PaPbP\|_{HS}}{\|P\|_{HS}} \\
&= \frac{\|P(Pa - aP)bP\|_{HS}}{\|P\|_{HS}} \\
&\leq \|b\| \frac{\|Pa - aP\|_{HS}}{\|P\|_{HS}} \\
&= \|b\| \left(\frac{\operatorname{Tr}(Paa^*P - PaPa^*P) + \operatorname{Tr}(Pa^*aP - Pa^*PaP)}{\operatorname{Tr}(P)} \right)^{1/2} \\
&= \|b\| \left(\operatorname{tr}_n(\phi(aa^*) - \phi(a)\phi(a^*)) + \operatorname{tr}_n(\phi(a^*a) - \phi(a^*)\phi(a)) \right)^{1/2}.
\end{aligned}
$$

$\square$

Recall that if τ is a tracial state on A then there is a "right regular" representation $\pi_\tau^{op} : A^{op} \to B(L^2(A, \tau))$ with the property that $\pi_\tau(A)' = \pi_\tau^{op}(A^{op})''$ and $\pi_\tau(A)'' = \pi_\tau^{op}(A^{op})'$. In particular, there is a natural $*$-homomorphism

$$A \odot A^{op} \to B(L^2(A, \tau)), a \otimes b \mapsto \pi_\tau(a)\pi_\tau^{op}(b).$$

Composing this representation with the vector state $x \mapsto \langle x\hat{1}, \hat{1}\rangle$, where $\hat{1} \in L^2(A, \tau)$ denotes the natural image of the unit of A, we get a positive linear functional μ_τ on $A \odot A^{op}$ which will play a distinguished role in what follows. The reader not familiar with this construction is advised to work out the case $A = M_n(\mathbb{C})$, $\tau = \operatorname{tr}_n$ as it is not only instructive but will also be used below.

We are now in a position to prove the main characterizations of amenable traces.

THEOREM 3.1.6 ([47]). *Let τ be a tracial state on A. Then the following are equivalent:*

(1) *τ is amenable.*
(2) *There exists a sequence of u.c.p. maps $\phi_n : A \to M_{k(n)}(\mathbb{C})$ such that $\|\phi_n(ab) - \phi_n(a)\phi_n(b)\|_2 \to 0$ and $\tau(a) = \lim_{n\to\infty} \operatorname{tr}_{k(n)} \circ \phi_n(a)$, for all $a, b \in A$.*
(3) *The positive linear functional $\mu_\tau : A \odot A^{op} \to \mathbb{C}$ is continuous with respect to the minimal tensor product norm (i.e. extends to a state on $A \otimes A^{op}$).*
(4) *The natural $*$-homomorphism $A \odot A^{op} \to B(L^2(A, \tau))$ is continuous with respect to the minimal tensor product norm (i.e. extends to a representation of $A \otimes A^{op}$).*
(5) *For any faithful representation $A \subset B(H)$ there exists a u.c.p. map $\Phi : B(H) \to \pi_\tau(A)''$ such that $\Phi(a) = \pi_\tau(a)$.[2]*

PROOF. (1) $\implies$ (2).[3] Let $A \subset B(H)$ be a faithful representation. Since τ is an amenable trace we can find a state ψ on $B(H)$ which extends τ and such

---

[2]The proof will show that this is also equivalent to knowing that *there exists* a faithful representation $A \subset B(H)$ and there exists a u.c.p. map $\Phi : B(H) \to \pi_\tau(A)''$ such that $\Phi(a) = \pi_\tau(a)$.

[3]The main idea in this implication comes directly from Connes' celebrated uniqueness theorem for the injective II_1-factor [23].

that $\psi(uTu^*) = \psi(T)$ for all unitaries $u \in A$ and operators $T \in B(H)$. Since the normal states on $B(H)$ are dense in the set of all states on $B(H)$ we can find a net of positive operators $h_\lambda \in \mathcal{T}$ such that $\mathrm{Tr}(h_\lambda T) \to \psi(T)$ for all $T \in B(H)$. Since $\psi(u^*Tu) = \psi(T)$ it follows that $\mathrm{Tr}(h_\lambda T) - \mathrm{Tr}((uh_\lambda u^*)T) \to 0$ for every $T \in B(H)$ and unitary $u \in A$. In other words, for every unitary $u \in A$ the net of trace class operators $h_\lambda - uh_\lambda u^*$ tends to zero in the weak topology (coming from $B(H)$). Hence, by the Hahn-Banach theorem, there are convex combinations which tend to zero in L^1-norm. In fact, taking finite direct sums (i.e. considering n-tuples $(u_1 h_\lambda u_1^* - h_\lambda, \ldots, u_n h_\lambda u_n^* - h_\lambda)$) one applies a similar argument to show that if $\mathfrak{F} \subset A$ is a finite set of unitaries then for every $\epsilon > 0$ we can find a positive trace class operator h such that $\mathrm{Tr}(h) = 1$, $|\mathrm{Tr}(uh) - \tau(u)| < \epsilon$ and $\|h - uhu^*\|_1 < \epsilon$ for all $u \in \mathfrak{F}$. Since finite rank operators are norm dense in the trace class operators we may further assume that h is finite rank with rational eigenvalues.

Applying Lemma 3.1.4 to bigger and bigger finite sets of unitaries and smaller and smaller ϵ's we can construct a sequence of u.c.p. maps $\phi_n : B(H) \to M_{k(n)}(\mathbb{C})$ such that $\mathrm{tr}_{k(n)}(\phi_n(u)) \to \tau(u)$ and $|\mathrm{tr}_{k(n)}(\phi_n(uu^*)) - \phi_n(u)\phi_n(u^*))| \to 0$ for every unitary u in a countable set with dense linear span in A. Of course, we may further assume that this set of unitaries is closed under the adjoint operation. From Lemma 3.1.5 it follows that for every unitary in this set and every $a \in A$ we have

$$\|\phi_n(ua) - \phi_n(u)\phi_n(a)\|_2 \to 0.$$

Since every element in A is a linear combination of four unitaries it follows that

$$\|\phi_n(ab) - \phi_n(a)\phi_n(b)\|_2 \to 0$$

and

$$\mathrm{tr}_{k(n)}(\phi_n(a)) \to \tau(a)$$

for all $a, b \in A$.

(2) $\Longrightarrow$ (3). We first note that it suffices to show that μ_τ is the weak-$*$ limit of *min*-continuous functionals. That is, if we can find a sequence of functionals $\sigma_n : A \otimes A^{op} \to \mathbb{C}$ such that $\sigma_n(a \otimes b) \to \mu_\tau(a \otimes b)$ for all $a \in A, b \in A^{op}$ then it will follow that any weak-$*$ cluster point of the σ_n's will be an extension of μ_τ to $A \otimes A^{op}$ (i.e. μ_τ is continuous with respect to the minimal tensor product norm).

So let $\phi_n : A \to M_{k(n)}(\mathbb{C})$ be a sequence of u.c.p. maps with the property that $\|\phi_n(ab) - \phi_n(a)\phi_n(b)\|_2 \to 0$ and $\tau(a) = \lim_{n \to \infty} \mathrm{tr}_{k(n)} \circ \phi_n(a)$, for all $a, b \in A$. Note we can also regard these maps as going from A^{op} to $M_{k(n)}(\mathbb{C})^{op}$ and they are still u.c.p. maps. To distinguish them, however, we let $\phi_n^{op} : A^{op} \to M_{k(n)}(\mathbb{C})^{op}$ be the "opposite" maps (though they are literally the same as ϕ_n as maps of linear spaces). Since u.c.p. maps behave well with respect to minimal tensor products we may consider the u.c.p. maps

$$\phi_n \otimes \phi_n^{op} : A \otimes A^{op} \to M_{k(n)}(\mathbb{C}) \otimes M_{k(n)}(\mathbb{C})^{op}.$$

All we need to show is that there is a state μ_n on $M_{k(n)}(\mathbb{C}) \otimes M_{k(n)}(\mathbb{C})^{op}$ such that $\mu_n \circ \phi_n \otimes \phi_n^{op} \to \mu_\tau$ in the weak-$*$ topology. In this picture it may not be so clear what the right state is but if we identify $M_{k(n)}(\mathbb{C}) \otimes M_{k(n)}(\mathbb{C})^{op}$ with $B(L^2(M_{k(n)}(\mathbb{C}), \mathrm{tr}_{k(n)}))$, $M_{k(n)}(\mathbb{C}) \otimes 1$ with the image of $M_{k(n)}(\mathbb{C})$ in the GNS representation with respect to the unique tracial state and $1 \otimes M_{k(n)}(\mathbb{C})^{op}$ with the commutant of $M_{k(n)}(\mathbb{C}) \subset B(L^2(M_{k(n)}(\mathbb{C}), \mathrm{tr}_{k(n)}))$ then it should be clear which state to pick. Let μ_n be the vector state on $B(L^2(M_{k(n)}(\mathbb{C}), \mathrm{tr}_{k(n)}))$ given by

$x \mapsto \langle x\hat{1}, \hat{1}\rangle$. Observe that

$$\mu_n(\phi_n \otimes \phi_n^{op}(a \otimes b)) = \langle \phi_n(a) J\phi_n(b^*)J\hat{1}, \hat{1}\rangle = \mathrm{tr}_{k(n)}(\phi_n(a)\phi_n(b)).$$

The Cauchy-Schwartz inequality shows that for all $x \in M_{k(n)}(\mathbb{C})$, $|\mathrm{tr}_{k(n)}(x)| \leq \|x\|_2$ and hence

$$|\mathrm{tr}_{k(n)}(\phi_n(a)\phi_n(b)) - \mathrm{tr}_{k(n)}(\phi_n(ab))| \to 0.$$

Thus we see that

$$\mu_n(\phi_n \otimes \phi_n^{op}(a \otimes b)) \to \tau(ab) = \langle \pi_\tau(a)\pi_\tau(b)\hat{1}, \hat{1}\rangle = \mu_\tau(a \otimes b),$$

for all $a, b \in A$.

(3) $\implies$ (4) follows from (the proof of) uniqueness of GNS representations. Indeed, on the norm dense $*$-subalgebra $A \odot A^{op} \subset A \otimes A^{op}$ we can construct a unitary operator which conjugates the representation $A \odot A^{op} \to B(l^2(A, \tau))$ to the GNS representation of $A \otimes A^{op}$ with respect to μ_τ.

(4) $\implies$ (5) uses an argument of Lance which we feel is the single most useful trick in the theory of tensor products. Since $A \otimes A^{op} \subset B(H) \otimes A^{op}$ we can extend the $*$-homomorphism $\pi_\tau \otimes \pi_\tau^{op} : A \otimes A^{op} \to B(L^2(A, \tau))$ to a completely positive map $\Phi : B(H) \otimes A^{op} \to B(L^2(A, \tau))$. Since $\Phi|_{A \otimes A^{op}}$ is a homomorphism it follows that $A \otimes A^{op}$ (and, in particular, $1 \otimes A^{op}$) is in the multiplicative domain of Φ. Hence, for every $T \in B(H)$, it follows that $\Phi(T \otimes 1) \in \Phi(1 \otimes A^{op})' = \pi_\tau^{op}(A^{op})' = \pi_\tau(A)''$.

(5) $\implies$ (1). Since A is contained in the multiplicative domain of Φ, it is easy to verify that $\varphi(T) = \langle \Phi(T)\hat{1}, \hat{1}\rangle$ defines a state on $B(H)$ which both extends τ and is invariant under the action of the unitary group of A on $B(H)$ and hence τ is an amenable trace. $\square$

Though the previous theorem will get the most use we should also point out a few other characterizations of amenable traces.

THEOREM 3.1.7. *Let $A \subset B(H)$ be in general position (i.e. $A \cap \mathcal{K}(H) = \{0\}$) and $\tau \in \mathrm{T}(A)$ be a tracial state. Then the following are equivalent:*

(1) *τ is an amenable trace.*
(2) *There exists a $*$-monomorphism $\sigma : \pi_\tau(A)'' \hookrightarrow R^\omega$ such that $\tau'' = \tau_\omega \circ \sigma$ and $\sigma \circ \pi_\tau : A \to R^\omega$ can be lifted to a u.c.p. map $\Phi : A \to l^\infty(R)$, where τ'' is the vector trace on $\pi_\tau(A)''$ induced by τ. (Kirchberg called these "liftable" traces in [47].)*
(3) *There exist finite rank projections $P_n \in B(H)$ (not necessarily increasing) such that*

$$\frac{\|aP_n - P_n a\|_{HS}}{\|P_n\|_{HS}} \to 0 \text{ and } \tau(a) = \lim_{n \to \infty} \frac{\langle aP_n, P_n\rangle_{HS}}{\langle P_n, P_n\rangle_{HS}},$$

for all $a \in A$. (This is Connes' "Følner" condition which he used to characterize the hyperfinite II_1-factor [23].)

PROOF. We first show that the second statement above is equivalent to the finite dimensional approximation property which is statement (2) in Theorem 3.1.6 and then we observe that statement (3) above is also equivalent to the finite dimensional approximation property.

(1) $\implies$ (2).[4] Assume there exists a sequence of u.c.p. maps $\phi_n : A \to M_{k(n)}(\mathbb{C})$ such that $\|\phi_n(ab) - \phi_n(a)\phi_n(b)\|_2 \to 0$ and $\tau(a) = \lim_{n\to\infty} \operatorname{tr}_{k(n)} \circ \phi_n(a)$, for all $a, b \in A$. Regard each matrix algebra $M_{k(n)}(\mathbb{C})$ as a subfactor of R and consider the direct sum u.c.p. map $\Phi : A \to l^\infty(R)$ given by

$$\Phi(a) = (\phi_n(a)).$$

It is not hard to see that composing with the quotient map $l^\infty(R) \to R^\omega$ yields a τ preserving $*$-homomorphism $A \to R^\omega$ (with u.c.p. lifting Φ). Finally one checks (essentially due to uniqueness of GNS representations) that the weak closure (in R^ω) of this $*$-homomorphism is isomorphic to $\pi_\tau(A)''$.

(2) $\implies$ (1). For this direction we have a couple of options. We prefer the approximation ideas (and these ideas will keep resurfacing throughout the paper) so our proof will use that technology.[5] So assume the existence of a u.c.p. map $\Phi : A \to l^\infty(R)$ such that composition with the quotient map $l^\infty(R) \to R^\omega$ yields a τ-preserving $*$-homomorphism $A \to R^\omega$. It suffices to show that if a finite set $\mathfrak{F} \subset A$ and $\epsilon > 0$ are given then we can find a u.c.p. map $\phi : A \to M_n(\mathbb{C})$ which is ϵ-multiplicative (in 2-norm) and almost recovers τ on the finite set $\mathfrak{F}$.

Let $\Phi_n : A \to R$ be the map $\Phi : A \to l^\infty(R) = \Pi_\mathbb{N}$ composed with the projection map from $l^\infty(R)$ onto the nth summand. Since Φ is multiplicative modulo the ideal

$$I_\omega = \{(x_n) : \lim_{n\to\omega} \|x_n\|_2 = 0\}$$

it follows that

$$\|\Phi_n(ab) - \Phi_n(a)\Phi_n(b)\|_2 \to 0 \text{ as } n \to \omega.$$

Hence for large enough n the maps Φ_n are almost 2-norm multiplicative on $\mathfrak{F}$ and recover τ. The proof is then completed by replacing R with a finite dimensional matrix algebra which is possible since R is hyperfinite.

That statement (3) above implies τ is amenable is quite simple since cutting by the projections P_n yields u.c.p. maps to matrix algebras which are asymptotically multiplicative and asymptotically recover τ. The converse is a consequence of Voiculescu's Theorem (version 2.2) since we assumed A is in general position. $\square$

3.2. Uniform amenable traces

Having seen that amenable traces are characterized by a finite dimensional approximation property we can now define smaller subsets of traces by demanding more of the approximants.

DEFINITION 3.2.1. A trace τ will be called *uniform amenable* if there exists a sequence of u.c.p. maps $\phi_n : A \to M_{k(n)}(\mathbb{C})$ such that $\|\phi_n(ab) - \phi_n(a)\phi_n(b)\|_2 \to 0$, for all $a, b \in A$, and

$$\|\tau - \operatorname{tr}_{k(n)} \circ \phi_n\|_{A^*} \to 0$$

where $\|\cdot\|_{A^*}$ is the natural norm on the dual Banach space of A. The set of all such traces will be denoted UAT(A).

---

[4]Though this is implication is well known to the experts and fairly straightforward, the reader unfamiliar with this type of argument is advised to nail down every detail as we will see several variations on this argument later on.

[5]A more elegant approach, left to the reader, would be to use statement (5) of Theorem 3.1.6. The idea is that if $A \subset B(H)$ and $\Phi : A \to l^\infty(R)$ is the assumed u.c.p. lifting then we may extend Φ to all $B(H)$, by the injectivity of $l^\infty(R)$, then pass to the quotient R^ω and, finally, compose with a conditional expectation $R^\omega \to \sigma(\pi_\tau(A)'')$.

Similar to the case of amenable traces, the space UAT(A) admits a number of nice characterizations. We thank Yasuyuki Kawahigashi, Sergei Neshveyev and Narutaka Ozawa for discussions in Oberwolfach which shortened our original proof of (1) $\Longrightarrow$ (2). Ozawa also added (6) to the list below and showed us the elegant proof.

THEOREM 3.2.2. *For a trace $\tau \in T(A)$, the following are equivalent:*
(1) $\tau \in \mathrm{UAT}(A)$.
(2) *There exist u.c.p. maps $\psi_n : A^{**} \to M_{k(n)}(\mathbb{C})$ such that for each free ultrafilter $\omega \in \beta\mathbb{N}\backslash\mathbb{N}$ we have*
$$\lim_{n\to\omega} \|\psi_n(xy) - \psi_n(x)\psi_n(y)\|_2 = 0 \text{ and } \lim_{n\to\omega} \mathrm{tr}_{k(n)} \circ \psi_n(x) = \tau^{**}(x),$$
for all $x,y \in A^{**}$, where τ^{**} is the normal trace on A^{**} induced by τ.
(3) *There exist u.c.p. maps $\psi_n : \pi_\tau(A)'' \to M_{k(n)}(\mathbb{C})$ such that for each free ultrafilter $\omega \in \beta\mathbb{N}\backslash\mathbb{N}$ we have*
$$\lim_{n\to\omega} \|\psi_n(xy) - \psi_n(x)\psi_n(y)\|_2 = 0 \text{ and } \lim_{n\to\omega} \mathrm{tr}_{k(n)} \circ \psi_n(x) = \tau''(x),$$
for all $x,y \in \pi_\tau(A)''$, where τ'' is the normal trace on $\pi_\tau(A)''$ induced by τ.
(4) *There exists a u.c.p. liftable, normal $*$-monomorphism $\sigma : \pi_\tau(A)'' \hookrightarrow R^\omega$.*
(5) *$\pi_\tau(A)''$ is hyperfinite.*
(6) *$\pi_\tau : A \to \pi_\tau(A)''$ is weakly nuclear (i.e. there exists u.c.p. maps $\phi_n : A \to M_{k(n)}(\mathbb{C})$, $\psi_n : M_{k(n)}(\mathbb{C}) \to \pi_\tau(A)''$ such that $\psi_n \circ \phi_n(a) \to \pi_\tau(a)$ in the σ-weak topology for every $a \in A$).*

PROOF. (1) $\Longrightarrow$ (2). Let $\phi_n : A \to M_{k(n)}(\mathbb{C})$ be 2-norm asymptotically multiplicative u.c.p. maps such that $\mathrm{tr}_{k(n)} \circ \phi_n \to \tau$ in the norm on A^*. Let $\phi_n^{**} : A^{**} \to M_{k(n)}(\mathbb{C})$ be the canonical normal extensions to the double dual. As is well known, and easily checked, ϕ_n^{**} are also u.c.p. maps. Hence $\mathrm{tr}_{k(n)} \circ \phi_n^{**}$ is a normal state on A^{**} and is in fact equal to the normal state on A^{**} induced by the functional $\mathrm{tr}_{k(n)} \circ \phi_n$ on A (they are both normal and agree on A). Since $\|\mathrm{tr}_{k(n)} \circ \phi_n - \tau\|_{A^*} \to 0$, it follows that $\mathrm{tr}_{k(n)} \circ \phi_n^{**}(x) \to \tau^{**}(x)$ for every $x \in A^{**}$. Hence, as in the proof of (1) $\Longrightarrow$ (2) from Theorem 3.1.7, we can construct a u.c.p. map $\Phi : A^{**} \to R^\omega$ such that $i)$ $\tau_\omega \circ \Phi = \tau^{**}$, $ii)$ $\Phi|_A$ is a $*$-homomorphism and $iii)$ Φ has a u.c.p. lifting $A^{**} \to l^\infty(R)$.[6] If we knew that Φ was a homomorphism then it would follow that $\lim_{n\to\omega} \|\phi_n^{**}(xy) - \phi_n^{**}(x)\phi_n^{**}(y)\|_2 = 0$ for all $x,y \in A^{**}$ and hence this is what we will show.

First note that Φ is normal: if $\{x_\lambda\} \subset A^{**}_{sa}$ is a norm bounded, increasing net of self adjoint elements with strong operator topology limit x then $\{\Phi(x_\lambda)\}$ is increasing up to $\Phi(x)$ (in the strong operator topology – i.e. 2-norm) since $\Phi(x_\lambda) \leq \Phi(x)$ and $\tau_\omega \circ (\Phi(x_\lambda)) = \tau^{**}(x_\lambda) \to \tau^{**}(x) = \tau_\omega \circ \Phi(x)$. It follows that Φ is continuous from the σ-weak topology on A^{**} to the σ-weak topology on R^ω (i.e. with respect to the weak-$*$ topologies coming from the preduals). Letting $\Psi : A^{**} \to R^\omega$ be the (weak-$*$ continuous) $*$-homomorphism which extends $\Phi|_A$ (and which exists by universality of A^{**}) it follows that $\Phi = \Psi$ since they are continuous and agree on A. Hence Φ is also multiplicative.

---
[6]As we have seen before, one first identifies each $M_{k(n)}(\mathbb{C})$ with a subfactor of R and then considers the direct sum map $A^{**} \to l^\infty(R)$, $x \mapsto (\phi_n^{**}(x))$, and finally composes with the quotient map to R^ω.

(2) $\implies$ (3). Assuming (2), we can use the maps ψ_n to construct a u.c.p. liftable, $*$-homomorphism $\sigma : A^{**} \to R^\omega$ such that $\tau_\omega \circ \sigma = \tau^{**}$. It follows that $\sigma(A^{**}) \cong \pi_\tau(A)''$ and, hence, $A^{**} \cong ker(\sigma) \oplus \pi_\tau(A)''$. Restricting the maps ψ_n to this non-unital copy of $\pi_\tau(A)''$ gives c.p. maps with the desired properties. Then applying Lemma 2.7 we can replace these nonunital maps with unital ones and we get (3).

(3) $\implies$ (4) is similar to the proof of (1) $\implies$ (2) from Theorem 3.1.7.

(4) $\implies$ (5). Identify $\pi_\tau(A)''$ with $\sigma(\pi_\tau(A)'') \subset R^\omega$ and let $\Phi : \pi_\tau(A)'' \to l^\infty(R)$ be a u.c.p. splitting and $E : R^\omega \to \pi_\tau(A)''$ be any conditional expectation. Now assume that $X \subset Y$ is an inclusion of operator systems and $\phi : X \to \pi_\tau(A)''$ is a u.c.p. map. Since $l^\infty(R)$ is injective we may extend the map $\Phi \circ \phi : X \to l^\infty(R)$ to all of Y. Composing this extension with the quotient map $l^\infty(R) \to R^\omega$ followed by the conditional expectation $E : R^\omega \to \pi_\tau(A)''$ yields a u.c.p. map $Y \to \pi_\tau(A)''$. This map extends ϕ since Φ was a lifting.

(5) $\implies$ (1) is a simple consequence of Lemmas 2.4 and 2.5 since hyperfinite von Neumann algebras contain weakly dense, finite dimensional subalgebras.

At this point we have shown that (1) - (5) are equivalent.

(5) $\implies$ (6) is trivial[7] and hence we are left to prove (6) $\implies$ (5). So assume that $\pi_\tau : A \to \pi_\tau(A)''$ is weakly nuclear and $\phi_n : A \to M_{k(n)}(\mathbb{C})$, $\psi_n : M_{k(n)}(\mathbb{C}) \to \pi_\tau(A)''$ are u.c.p. maps whose composition converges to π_τ in the point-σ-weak topology. Using these maps it is not hard to see that the canonical homomorphism $A \odot \pi_\tau(A)' \to B(L^2(A,\tau))$, $a \otimes x \mapsto \pi_\tau(a)x$, is continuous with respect to the minimal tensor product norm. (Use the fact that the natural map on the maximal tensor product approximately factorizes through $M_{k(n)} \otimes_{max} \pi_\tau(A)' = M_{k(n)} \otimes \pi_\tau(A)'$ and hence factors through the minimal tensor product.) As in the proof of (4) $\implies$ (5) from Theorem 3.1.6, it follows that there exists a conditional expectation $B(L^2(A,\tau)) \to \pi_\tau(A)'$ and hence $\pi_\tau(A)'$ is injective. This implies that $\pi_\tau(A)''$ is also injective and the proof is complete. $\square$

Note that part (3) in the previous theorem could be used as an abstract (i.e. representation free) definition of quasidiagonality, analogous to Voiculescu's abstract characterization, in the setting of tracial von Neumann algebras. The equivalence of (3) and (5) would then say that quasidiagonality is equivalent to hyperfiniteness. Note that this is in stark contrast to the C*-case where Dadarlat has constructed non-nuclear Popa – hence quasidiagonal – algebras [**26**].

REMARK 3.2.3. Though we prefer to state all approximation properties in this paper in terms of matrix algebras there is one possible advantage to using general finite dimensional approximants. Namely, if τ is a uniform amenable trace on A then there exist finite dimensional C*-algebras B_n, traces $\tau_n \in \mathrm{T}(B_n)$ and u.c.p. maps $\phi_n : A \to B_n$ such that $\|\phi_n(ab) - \phi_n(a)\phi_n(b)\|_{2,\tau_n} \to 0$ for all $a,b \in A$ and (here is the advantage)
$$\tau = \tau_n \circ \phi_n$$

---

[7]Use the fact that finite dimensional subalgebras are weakly dense in hyperfinite von Neumann algebras. Perhaps we should also remark that there is no difference between only considering matrix algebras, as opposed to general finite dimensional algebras, since every finite dimensional algebra unitally embeds into a matrix algebra and there will always be a conditional expectation from the larger matrix algebra back onto the original finite dimensional algebra.

for all $n \in \mathbb{N}$. The proof of this uses the fact that $\pi_\tau(A)''$ is hyperfinite and hence there are τ-preserving conditional expectations onto the finite dimensional subalgebras of $\pi_\tau(A)''$.

3.3. Quasidiagonal traces

In the previous section we strengthened the approximation property of amenable traces by requiring norm convergence, as opposed to weak-$*$ convergence, to the trace τ. We now define a subset of traces by requiring a stronger notion of asymptotic multiplicativity. Since we have two choices for convergence to τ this actually leads to two subsets of AT(A).

DEFINITION 3.3.1. We say a trace $\tau \in \mathrm{T}(A)$ is *quasidiagonal* if there exist u.c.p. maps $\phi_n : A \to M_{k(n)}(\mathbb{C})$ such that $\mathrm{tr}_{k(n)} \circ \phi_n \to \tau$ in the weak-$*$ topology and
$$\|\phi_n(ab) - \phi_n(a)\phi_n(b)\| \to 0$$
for all $a, b \in A$.[8] A trace $\tau \in \mathrm{T}(A)$ will be called *uniform quasidiagonal* if one can further arrange that
$$\|\tau - \mathrm{tr}_{k(n)} \circ \phi_n\|_{A^*} \to 0.$$
These sets of traces will be denoted AT(A)$_{\mathrm{QD}}$ and UAT(A)$_{\mathrm{QD}}$, respectively.

These sets are far more mysterious than the amenable traces.[9] For example, we don't know whether every amenable trace is quasidiagonal. Though we doubt that this is the case, at the moment we have no good way of distinguishing the two sets. (Compare with Blackadar and Kirchberg's question [13]: Is every nuclear, stably finite C*-algebra necessarily quasidiagonal?) If one starts with a quasidiagonal C*-algebra, however, then it turns out that AT(A)$_{\mathrm{QD}}$ is *precisely* the set of traces which can be encoded in the definition of quasidiagonality. Recall that, by definition, a C*-algebra A is quasidiagonal if there exists a faithful representation $\pi : A \to B(H)$ such that one can find an increasing sequence of finite rank projections $P_1 \leq P_2 \leq \ldots$ with the property that $\|\pi(a)P_n - P_n\pi(a)\| \to 0$ for all $a \in A$ and $P_n \to 1_H$ in the strong operator topology. (This is not the right definition for non-separable algebras.)

PROPOSITION 3.3.2. *Let $A \subset B(H)$ be in general position (i.e. $A \cap \mathcal{K}(H) = 0$). If A is quasidiagonal then* AT(A)$_{\mathrm{QD}} \neq \emptyset$ *and, moreover, there exists an increasing sequence of finite rank projections $P_1 \leq P_2 \leq \ldots$, converging strongly to the identity, which asymptotically commutes (in norm) with every element in A and such that for each $\tau \in$ AT(A)$_{\mathrm{QD}}$ there exists a subsequence $\{n_k\}$ such that*
$$\frac{\langle aP_{n_k}, P_{n_k}\rangle_{HS}}{\langle P_{n_k}, P_{n_k}\rangle_{HS}} \to \tau(a), \ as \ k \to \infty,$$
for all $a \in A$.

PROOF. That AT(A)$_{\mathrm{QD}} \neq \emptyset$ is well known (cf. [78, 2.4], [17, Proposition 6.1]), but we remind the reader of the proof. If $\pi : A \to B(H)$ and $P_1 \leq P_2 \leq \ldots$ are as in the definition of quasidiagonality then one defines u.c.p. maps by
$$\phi_n(a) = P_n\pi(a)P_n.$$

---

[8]This terminology is inspired by Voiculescu's abstract characterization of quasidiagonal C*-algebras: A is quasidiagonal if and only there exist u.c.p. maps $\phi_n : A \to M_{k(n)}(\mathbb{C})$ such that $\|\phi_n(ab) - \phi_n(a)\phi_n(b)\| \to 0$ and $\|a\| = \lim \|\phi_n(a)\|$ for all $a, b \in A$.

[9]But also more important to the classification program.

The asymptotic commutativity of P_n ensures that ϕ_n are asymptotically multiplicative in norm. Note also that $P_n B(H) P_n \cong M_{rank(P_n)}(\mathbb{C})$. Finally, a straightforward calculation shows that any weak-$*$ cluster point of the sequence of states $\{\mathrm{tr}_{rank(P_n)} \circ \phi_n\}$ is necessarily a tracial state and hence $\mathrm{AT}(A)_{\mathrm{QD}} \neq \emptyset$. (Note that we used the assumption that A is unital here; in the non-unital case it can happen that $\mathrm{T}(A) = \emptyset$ – e.g. the suspension of a Cuntz algebra).

To prove the rest of the proposition we claim that it suffices to show that for every finite set $\mathfrak{F} \subset A$, finite dimensional subspace $X \subset H$, $\varepsilon > 0$ and trace $\tau \in \mathrm{AT}(A)_{\mathrm{QD}}$ there exists a finite rank projection $P \in B(H)$ such that

(1) $\|[a, P]\| < \varepsilon$ for all $a \in \mathfrak{F}$.
(2) $P(x) = x$ for all $x \in X$.
(3) $|\frac{\langle aP, P \rangle_{HS}}{\langle P, P \rangle_{HS}} - \tau(a)| < \varepsilon$ for all $a \in \mathfrak{F}$.

Assume for the moment that we were able to prove this local version. Then, if $\{a_n\} \subset A$ is a sequence which is dense in the unit ball of A and $\{\tau_j\}$ is any sequence of traces in $\mathrm{AT}(A)_{\mathrm{QD}}$ we could apply the above local approximation property to construct a sequence $P_1 \leq P_2 \leq \ldots$ which was converging strongly to the identity, asymptotically commuting in norm with A and such that

$$|\frac{\langle a_i P_n, P_n \rangle_{HS}}{\langle P_n, P_n \rangle_{HS}} - \tau_n(a_i)| < 1/n$$

for all $n \in \mathbb{N}$ and $1 \leq i \leq n$. Since A is separable, the weak-$*$ topology on $\mathrm{T}(A)$ is metrizable and hence we can always find a sequence of traces $\{\tau_j\} \subset \mathrm{AT}(A)_{\mathrm{QD}}$ such that there exists a weak-$*$ dense subset $Y \subset \mathrm{AT}(A)_{\mathrm{QD}}$ with the property that every element of Y appears infinitely many times in the sequence $\{\tau_j\}$. The sequence of projections associated with such a sequence of traces will have all the properties asserted in the statement of the proposition. Hence it suffices to prove the local statement in the first paragraph of the proof.

The required local statement is now a consequence of Voiculescu's Theorem (version 2.3) and a little trickery. Let $\tau \in \mathrm{AT}(A)_{\mathrm{QD}}$ be arbitrary. Since A is quasidiagonal, by Lemma 2.6 we can find a sequence of u.c.p. maps $\phi_n : A \to M_{k(n)}(\mathbb{C})$ which are asymptotically multiplicative, asymptotically isometric and such that $\mathrm{tr}_{k(n)} \circ \phi_n \to \tau$ in the weak-$*$ topology. If a finite set $\mathfrak{F} \subset A$ and $\epsilon > 0$ are given then, by passing to a subsequence if necessary, we may assume that $\|\phi_n(ab) - \phi_n(a)\phi_n(b)\| < \varepsilon$ and $|\mathrm{tr}_{k(n)} \circ \phi_n(a) - \tau(a)| < \varepsilon$ for all n and for all $a, b \in \mathfrak{F}$. Letting $K = \oplus_n \mathbb{C}^{k(n)}$ and $\Phi = \oplus_n \phi_n : A \to B(K)$ we have that Φ is a faithful $*$-homomorphism modulo the compacts. Hence we can find a unitary operator $U : K \to H$ such that $U\Phi(a)U^*$ is nearly equal (in norm) to a, for all $a \in \mathfrak{F}$. Hence, if $Q_s \in B(K)$ is the orthogonal projection onto $\oplus_1^s \mathbb{C}^{k(n)}$ we have that $\|[UQ_s U^*, a]\|$ is small for all $s \in \mathbb{N}$ and for all $a \in A$. Moreover, compressing $a \in \mathfrak{F}$ to the range of $UQ_s U^*$ will almost recover the trace τ (for all s). Finally, if $X \subset H$ is any finite dimensional subspace then X will almost be contained in the range of $UQ_s U^*$ for sufficiently large s and hence a tiny (norm) perturbation of a sufficiently large $UQ_s U^*$ will actually be the identity on X (and still almost commute with $\mathfrak{F}$ and almost recover the trace τ). $\square$

It is natural to wonder whether or not every trace on a quasidiagonal C*-algebra arises as above; i.e. whether or not $\mathrm{T}(A) = \mathrm{AT}(A)_{\mathrm{QD}}$ when A is quasidiagonal. This is the case for $A = C^*(\mathbb{F}_\infty)$ (the universal C*-algebra of a countably generated free group) if and only if Connes' embedding problem has an affirmative answer (see Proposition 6.3.5). Unfortunately this is not the case in general; there exist (exact) quasidiagonal C*-algebras such that $T(A) \neq AT(A)_{QD}$ (see Corollary 4.3.8).

3.4. Locally finite dimensional traces

There is at least one more natural set of traces which one may define – those arising as limits of finite dimensional traces (i.e. limits of traces whose GNS representation is finite dimensional). Though very natural, this set of traces would be empty for every infinite dimensional, simple C*-algebra (or any other C*-algebra which has no finite dimensional representations). However it turns out that there is another possibility for strengthening the definition of quasidiagonal trace which contains a number of non-trivial examples, though it is more technical to formulate. Recall that if $\phi : A \to B$ is a u.c.p. map then we denote the multiplicative domain of ϕ by A_ϕ.

DEFINITION 3.4.1. A trace τ will be called *locally finite dimensional* if there exist u.c.p. maps $\phi_n : A \to M_{k(n)}(\mathbb{C})$ such that $\mathrm{tr}_{k(n)} \circ \phi_n \to \tau$ in the weak-$*$ topology and
$$d(a, A_{\phi_n}) \to 0$$
for all $a \in A$.[10] A trace $\tau \in T(A)$ will be called *uniform locally finite dimensional* if one can further arrange that
$$\|\tau - \mathrm{tr}_{k(n)} \circ \phi_n\|_{A^*} \to 0.$$
These sets of traces will be denoted $AT(A)_{LFD}$ and $UAT(A)_{LFD}$, respectively.

Note that being locally finite dimensional is stronger than quasidiagonality of a trace. Indeed, if τ is locally finite dimensional, with $\phi_n : A \to M_{k(n)}(\mathbb{C})$ as in the definition, and $a, b \in A$ are given then we have
$$\phi_n(ab) \approx \phi_n(a_n b_n) = \phi_n(a_n)\phi_n(b_n) \approx \phi_n(a)\phi_n(b),$$
where $a_n, b_n \in A_{\phi_n}$ are chosen so that $\|a - a_n\|, \|b - b_n\| \to 0$.

At this point it is hard to motivate this definition but the interested reader may find it fun to prove that every trace on an AF algebra is uniform locally finite dimensional. (Even though there are plenty of simple, infinite dimensional AF algebras where limits of finite dimensional traces would be impossible.) In fact, we will see many more examples as well as observe that these traces are intimately related to future progress in the classification program. However, we having nothing more to say about this definition at the moment.

3.5. Miscellaneous remarks and permanence properties

A few remarks about all of these sets are probably in order. First note that we have the following inclusions:

$$\begin{array}{ccccccc} T(A) & \supset & AT(A) & \supset & AT(A)_{QD} & \supset & AT(A)_{LFD} \\ & & \cup & & \cup & & \cup \\ & & UAT(A) & \supset & UAT(A)_{QD} & \supset & UAT(A)_{LFD} \end{array}$$

A natural question to ask is whether or not any of these inclusions are proper. We will see that all of the vertical inclusions are proper in general, but for large

---

[10]$d(a, A_{\phi_n}) \to 0$ means there exist $a_n \in A_{\phi_n}$ such that $\|a - a_n\| \to 0$.

classes of C*-algebras (e.g. exact C*-algebras) one has the equalities $\mathrm{AT}(A) = \mathrm{UAT}(A)$ and $\mathrm{AT}(A)_{\mathrm{QD}} = \mathrm{UAT}(A)_{\mathrm{QD}}$ (see Theorem 4.3.3).[11] While it is not hard to give examples showing $T(A) \neq \mathrm{AT}(A)$ (see Proposition 4.1.1) we have not yet found an example which shows that the set of amenable traces can be strictly larger than the set of quasidiagonal traces.[12] We believe such examples should exist but it is not clear at this point where to look. Similarly, the relation between the quasidiagonal and the locally finite dimensional traces is not yet well understood.

Another natural question is whether or not any of these sets are convex and/or weak-$*$ closed subsets of $T(A)$.

PROPOSITION 3.5.1. *The sets* $\mathrm{AT}(A)$, $\mathrm{AT}(A)_{\mathrm{QD}}$, $\mathrm{UAT}(A)$ *and* $\mathrm{UAT}(A)_{\mathrm{QD}}$ *are all convex. Both* $\mathrm{AT}(A)$ *and* $\mathrm{AT}(A)_{\mathrm{QD}}$ *are closed in the weak-$*$ topology while the sets* $\mathrm{UAT}(A)$ *and* $\mathrm{UAT}(A)_{\mathrm{QD}}$ *are closed in norm (i.e. the norm on A^*) and thus, by the Hahn-Banach theorem, closed in the weak topology coming from A^{**}.*

PROOF. Since $\mathrm{AT}(A)_{\mathrm{QD}}$ and $\mathrm{AT}(A)$ (resp. $\mathrm{UAT}(A)$ and $\mathrm{UAT}(A)_{\mathrm{QD}}$) are defined via weak-$*$ convergence (resp. norm convergence), it is easy to see that these sets are closed in this topology.

The proof of the convexity assertion is a reasonable exercise. However, we will need little perturbations of this argument several times so perhaps it is worthwhile pointing out the main things to keep in mind for future arguments.[13] We only give the proof of convexity for $\mathrm{UAT}(A)$ as all other cases use similar arguments. So assume $\tau_1, \tau_2 \in \mathrm{UAT}(A)$, $0 < \lambda < 1$ and $\tau = \lambda \tau_1 + (1-\lambda)\tau_2$. If we fix a finite set $\mathfrak{F} \subset A$ and $\epsilon > 0$ then it is possible to find u.c.p. maps $\phi_i : A \to M_{n(i)}(\mathbb{C})$, $i = 1, 2$, such that

$$\|\phi_i(ab) - \phi_i(a)\phi_i(b)\|_{2,\mathrm{tr}_{n(i)}} < \epsilon$$

for all $a, b \in A$ and

$$\|\tau_i - \mathrm{tr}_{n(i)} \circ \phi_i\|_{A^*} < \epsilon.$$

We now consider the finite dimensional algebra

$$B = M_{n(1)}(\mathbb{C}) \oplus M_{n(2)}(\mathbb{C})$$

and the tracial state

$$\gamma = \lambda \mathrm{tr}_{n(1)} \oplus (1-\lambda) \mathrm{tr}_{n(2)}.$$

Let $\phi = \phi_1 \oplus \phi_2 : A \to B$ and some straightforward calculations show that $\|\phi(ab) - \phi(a)\phi(b)\|^2_{2,\gamma}$ is equal to

$$\lambda \|\phi_1(ab) - \phi_1(a)\phi_1(b)\|^2_{2,\mathrm{tr}_{n(1)}} + (1-\lambda)\|\phi_2(ab) - \phi_2(a)\phi_2(b)\|^2_{2,\mathrm{tr}_{n(2)}}$$

---

[11]We don't know if every exact C*-algebra satisfies the equation $\mathrm{AT}(A)_{\mathrm{LFD}} = \mathrm{UAT}(A)_{\mathrm{LFD}}$ and this turns out to be a very important open question (see Section 6.1).

[12]Even for nuclear C*-algebras A it appears to be very difficult to decide whether or not one always has $\mathrm{AT}(A) = \mathrm{AT}(A)_{\mathrm{QD}}$. If every nuclear C*-algebra satisfies the equation $\mathrm{AT}(A) = \mathrm{AT}(A)_{\mathrm{QD}}$ then every nuclear C*-algebra with a faithful trace would be quasidiagonal – in particular, Rosenberg's conjecture that $C^*_r(\Gamma)$ is quasidiagonal, for every discrete amenable group Γ, would follow. On the other hand, nuclear examples showing $\mathrm{AT}(A) \neq \mathrm{AT}(A)_{\mathrm{QD}}$ would likely have significant consequences for Elliott's classification program.

[13]One of the irritating aspects of our various definitions of approximation is that we have required the use of matrix algebras as opposed to general finite dimensional C*-algebras. This is mostly because it makes the paper easier to write! There is no loss of generality in requiring matrix algebras, but it is sometimes convenient in proofs to use general finite dimensional algebras. The argument we are about to sketch allows one to pass freely from one setting to another.

and
$$\|\tau - \gamma \circ \phi\|_{A^*} \leq \lambda \|\tau_1 - \operatorname{tr}_{n(1)} \circ \phi_1\|_{A^*} + (1-\lambda)\|\tau_2 - \operatorname{tr}_{n(2)} \circ \phi_2\|_{A^*}.$$

It follows that τ has the right approximation properties if we allow general finite dimensional C*-algebras. Hence we can complete the proof by appealing to Lemma 2.5 to replace (B, γ) with a full matrix algebra. □

Evidently AT(A)$_{\text{LFD}}$ (resp. UAT(A)$_{\text{LFD}}$) is also weak-* (resp. norm) closed but we don't know whether or not it is convex.

We will soon see that if Γ is a non-amenable, residually finite, discrete group then the canonical trace on $C^*(\Gamma)$ (which gives the left regular representation in the GNS construction) is always a weak-* limit of (uniform locally) finite dimensional traces, but is not itself a uniform amenable trace. Thus, the sets UAT(A), UAT(A)$_{\text{QD}}$ and UAT(A)$_{\text{LFD}}$ need not be weak-* closed in general.

Finally, one may wonder if any of these sets define a face in T(A). We are not sure about the other four sets, but this is the case for the (uniformly) amenable traces.

The proofs are a simple consequence of Theorems 3.1.6 and 3.2.2 and the following fact.

LEMMA 3.5.2. *Assume $\tau, \gamma \in$ T(A) and there exists $0 < s < 1$ such that $s\tau \leq \gamma$.[14] Then there exists a projection $P \in \pi_\gamma(A)'$ such that $a \mapsto P\pi_\gamma(a)$ is unitarily equivalent to π_τ.*

PROOF. By [61, Proposition 3.3.5] we can find a positive element $y \in \pi_\gamma(A)'$ such that
$$\tau(a) = \langle \pi_\gamma(a)y\hat{1}, \hat{1} \rangle,$$
for all $a \in A$. Uniqueness of GNS representations implies that the projection onto the closure of the subspace $\{\pi_\gamma(a)y\hat{1} : a \in A\}$ will do the trick. □

That AT(A) is a face was first observed by Kirchberg.

PROPOSITION 3.5.3. (cf. [47, Lemma 3.4]) *AT(A) is a face in T(A).*

PROOF. If $\tau_1, \tau_2 \in$ T(A), $0 < s < 1$ are given and $\gamma = s\tau_1 + (1-s)\tau_2 \in$ AT(A) then we can find projections $P_i \in \pi_\gamma(A)''$ such that $a \mapsto P_i \pi_\gamma(a)$ is unitarily equivalent to π_{τ_i} and (by amenability) we can find a u.c.p. map $\Phi : B(H) \to \pi_\gamma(A)''$ such that $\Phi(a) = \pi_\gamma(a)$ (for any fixed inclusion $A \subset B(H)$). Hence we can reapply condition (5) in Theorem 3.1.6 to deduce amenability of each τ_i by first taking the u.c.p. map Φ and composing with compression by P_i (identifying the latter map with the GNS representation coming from τ_i). □

PROPOSITION 3.5.4. *UAT(A) is a face in T(A).*

PROOF. The proof is very similar to the previous proposition. We use Theorem 3.2.2 and the fact that all von Neumann subalgebras of a finite, hyperfinite von Neumann algebra are again hyperfinite. □

We now wish to discuss a few permanence properties of the traces we have defined. We begin with two easy situations. In the first we consider restrictions of traces to subalgebras. The proof of the following proposition is a very simple exercise – just take the finite dimensional approximating maps which are given on

---

[14]This means $s\tau(a^*a) \leq \gamma(a^*a)$ for all $a \in A$.

the larger algebra and restrict them to the smaller algebra. Note that this obvious procedure does not work when considering locally finite dimensional traces as it is not clear that the multiplicative domain of a u.c.p. map defined on the big algebra would have to intersect the smaller algebra.[15]

PROPOSITION 3.5.5 (Restriction to subalgebras). *Assume $1_A \in B \subset A$ is a C^*-subalgebra and $\tau \in \mathrm{T}(A)$. If τ is amenable, uniform amenable, quasidiagonal or uniform quasidiagonal then $\tau|_B$ enjoys the same approximation property.*

Another trivial fact, again left to the reader, is that nice traces defined on quotients always lift to nice traces. In this situation, even the locally finite dimensional traces behave well.

PROPOSITION 3.5.6 (Lifting from a quotient). *If $I \triangleleft A$ is an ideal and $\tau \in \mathrm{T}(A/I)$ is a trace enjoying any one of the six approximation properties defined in these notes then the trace on A gotten by composing τ with the quotient map $A \to A/I$ inherits the same approximation property.*

A marginally less trivial situation is that of tensor products.

PROPOSITION 3.5.7. *Assume $\tau \in \mathrm{T}(A)$ and $\gamma \in \mathrm{T}(B)$.*
 (1) *If $\tau \in \mathrm{AT}(A)$ (resp. $\tau \in \mathrm{UAT}(A)$) and $\gamma \in \mathrm{AT}(B)$ (resp. $\gamma \in \mathrm{UAT}(B)$) then $\tau \otimes \gamma \in \mathrm{AT}(A \otimes B)$ (resp. $\tau \otimes \gamma \in \mathrm{UAT}(A \otimes B)$).*
 (2) *If $\tau \in \mathrm{AT}(A)_{\mathrm{QD}}$ (resp. $\tau \in \mathrm{UAT}(A)_{\mathrm{QD}}$) and $\gamma \in \mathrm{AT}(B)_{\mathrm{QD}}$ (resp. $\gamma \in \mathrm{UAT}(B)_{\mathrm{QD}}$) then $\tau \otimes \gamma \in \mathrm{AT}(A \otimes B)_{\mathrm{QD}}$ (resp. $\tau \otimes \gamma \in \mathrm{UAT}(A \otimes B)_{\mathrm{QD}}$).*
 (3) *If $\tau \in \mathrm{AT}(A)_{\mathrm{LFD}}$ (resp. $\tau \in \mathrm{UAT}(A)_{\mathrm{LFD}}$) and $\gamma \in \mathrm{AT}(B)_{\mathrm{LFD}}$ (resp. $\gamma \in \mathrm{UAT}(B)_{\mathrm{LFD}}$) then $\tau \otimes \gamma \in \mathrm{AT}(A \otimes B)_{LFD}$ (resp. $\tau \otimes \gamma \in \mathrm{UAT}(A \otimes B)_{LFD}$).*

PROOF. We think the details of the proof are best left as an exercise. One should recall, however, that one of the nice things about minimal tensor products and completely bounded maps is that if $\phi: A \to C$ and $\psi: B \to D$ are completely bounded maps (e.g. linear functionals – which, together with the case of u.c.p. maps, is what one must handle in the proof of the present proposition) then there is a well defined completely bounded map $\phi \otimes \psi : A \otimes B \to C \otimes D$ such that

$$\|\phi \otimes \psi\|_{CB} = \|\phi\|_{CB} \|\psi\|_{CB}.$$

One can refer to any book on operator space theory for this fundamental fact.

Given this fact all six cases are handled in the same fashion. Take u.c.p. maps on A and B with appropriate approximation properties and consider the tensor product of these maps. □

Determining when a trace with nice approximation properties which happens to vanish on an ideal descends to a nice trace on the quotient is sometimes easy, sometimes hard and sometimes not at all clear.[16]

PROPOSITION 3.5.8 (Passage to a quotient). *Let $\tau \in \mathrm{T}(A)$ be given, $I = ker(\pi_\tau(A))$, $B = A/I \cong \pi_\tau(A)$ and $\dot\tau \in \mathrm{T}(B)$ be the (faithful, vector) trace induced by τ.*

---

[15]We are not claiming that the restriction of a locally finite dimensional trace to a subalgebra need not be locally finite dimensional! Indeed, we don't know whether or not this is true. We are only claiming that the obvious proof doesn't seem to work.

[16]For example, it is not clear what happens with the locally finite dimensional traces.

(1) *If $\tau \in \mathrm{AT}(A)$ and there exists a contractive c.p. (c.c.p.) splitting $\Phi : B \to A$ of the quotient map $A \to B$ then $\dot\tau \in \mathrm{AT}(B)$.*
(2) *If $\tau \in \mathrm{UAT}(A)$ then $\dot\tau \in \mathrm{UAT}(B)$.*
(3) *If $\tau \in \mathrm{AT}(A)_{\mathrm{QD}}$, there exists a c.c.p. splitting $\Phi : B \to A$ of the quotient map $A \to B$ and the extension*
$$0 \to I \to A \to B \to 0$$
is quasidiagonal then $\dot\tau \in \mathrm{AT}(B)_{\mathrm{QD}}$.[17]
(4) *If $\tau \in \mathrm{UAT}(A)_{\mathrm{QD}}$, there exists a c.c.p. splitting $\Phi : B \to A$ of the quotient map $A \to B$ and the extension*
$$0 \to I \to A \to B \to 0$$
is quasidiagonal then $\dot\tau \in \mathrm{UAT}(B)_{\mathrm{QD}}$.

PROOF. Though none of these assertions are particularly difficult note that only the second one is obvious since the fifth statement in Theorem 3.2.2 asserts that the *image* of the GNS representation completely determines when a trace is uniform amenable (and the images of the GNS representations of (A, τ) and $(B, \dot\tau)$ are canonically unitarily equivalent).

For the proof of (1) we will apply the fifth condition from Theorem 3.1.6. Indeed, by assumption, if $A \subset B(H)$ and $B \subset B(K)$ are faithful representations then we can find a u.c.p. map
$$\Psi : B(H) \to \pi_\tau(A)'' = \pi_{\dot\tau}(B)''$$
such that $\Psi(a) = \pi_\tau(a)$ for all $a \in A$. By Arveson's extension theorem we may assume that the u.c.p. splitting $\Phi : B \to A$ is actually defined on all of $B(K)$ and takes values in $B(H)$. One readily verifies that the u.c.p. map $\Psi \circ \Phi : B(K) \to \pi_{\dot\tau}(B)''$ also satisfies condition (5) in Theorem 3.1.6 and hence $\dot\tau \in \mathrm{AT}(B)$.

The proofs of (3) and (4) are quite similar so we only give the proof of (4). Since $0 \to I \to A \to B \to 0$ is a quasidiagonal extension we can find a quasicentral approximate unit of projections, say $\{p_n\} \subset I$. If we consider the sequence of c.p. splittings
$$\Phi_n : B \to A, \quad \Phi_n(b) = (1 - p_n)\Phi(b)(1 - p_n)$$
then it is well known that these maps will be asymptotically multiplicative in norm.[18] Hence if a finite set $\mathfrak{F} \subset B$ and $\epsilon > 0$ are given then we can find n large enough that Φ_n is as close to multiplicative on $\mathfrak{F}$ as one desires. Then one can apply the definition of uniform quasidiagonal trace to the finite set $\Phi_n(\mathfrak{F}) \subset A$ to find a u.c.p. map $\phi : A \to M_n(\mathbb{C})$ which is almost multiplicative on $\Phi_n(\mathfrak{F})$ and almost recaptures τ (in the uniform norm) after composing with the trace on matrices. It is a simple exercise to check that $\phi \circ \Phi_n : B \to M_n(\mathbb{C})$ is almost multiplicative on $\mathfrak{F}$ and that
$$\|\mathrm{tr}_n \circ \phi \circ \Phi_n - \dot\tau\|_{B^*} \leq \|\mathrm{tr}_n \circ \phi - \tau\|_{A^*}.$$
Of course we really should have unital maps instead of just contractive c.p. maps but Lemma 2.7 allows us to circumvent this technicality and thus we deduce that $\dot\tau$ is uniform quasidiagonal on B. □

---

[17] An extension $0 \to I \to A \to B \to 0$ is called quasidiagonal if there exists an approximate unit $\{p_n\} \subset I$ which (a) is quasicentral in A and (b) each p_n is a projection.

[18] Since $\|\Phi_n(xy) - \Phi_n(x)\Phi_n(y)\| \leq \|(1-p_n)[\Phi(xy)-\Phi(x)\Phi(y)](1-p_n)\| + \|[p_n, \Phi(x)]\|\|y\|$ and the latter two norms tend to zero since $\{p_n\}$ is an approximate unit and quasicentral, respectively.

REMARK 3.5.9. In some instances we don't actually need the full power of a c.c.p. splitting $\Phi : B \to A$ – it may suffice to just have *local liftability* which asserts that for each finite dimensional operator system $X \subset B$ there should exist a u.c.p. map $\Phi_X : X \to A$ which lifts X. For example with this hypothesis the conclusions in (1) and (3) above still hold with a little more work. For (1) the idea of the proof is the same except one has to take a limit of maps using the fact that the set of u.c.p. maps between two von Neumann algebras is compact in the point ultraweak topology. For (3) the proof is again quite similar but one must apply Arveson's extension theorem at one point to extend maps which are only defined on finite dimensional operator subsystems of B to all of B. Since we won't need these results we leave the details to the interested reader. Finally we recall that the Effros-Haagerup lifting theorem asserts that local liftability is equivalent to knowing that for every C*-algebra D the sequence

$$0 \to I \otimes D \to A \otimes D \to B \otimes D \to 0$$

is exact (cf. [**28**]).

Though we will try our hardest to avoid non-unital C*-algebras they are occasionally forced upon us. Hence a few words about unitizations and amenable traces on ideals are probably in order. However, we don't want to spend time systematically studying this nonunital setting so we only present the results that we will need.

Assume that τ is a tracial state on a *nonunital* C*-algebra A and let $\tilde{A}$ be the unitization of A and $\tilde{\tau}$ denote the unique extension of τ to a tracial state on $\tilde{A}$. When trying to extend the six definitions of traces given in the previous sections one is faced with two natural possibilities. On the one hand it would be very natural to simply replace u.c.p. maps with c.c.p. maps, mimic the previous approximation properties and take these as the nonunital definitions. On the other hand it would also be natural to just define your way out of this issue by declaring that τ is quasidiagonal, for example, if $\tilde{\tau}$ is a quasidiagonal trace on $\tilde{A}$. Not surprisingly, in all six cases these two approaches yield the same sets of traces on A.

PROPOSITION 3.5.10 (Traces on nonunital C*-algebras). *If A is nonunital, $\tau \in \mathrm{T}(A)$, $\tilde{A}$ is the unitization of A and $\tilde{\tau} \in \mathrm{T}(\tilde{A})$ is the unique extension then the following statements hold.*

(1) *$\tilde{\tau}$ is amenable (resp. uniform amenable) if and only if there exist c.c.p. maps $\phi_n : A \to M_{k(n)}(\mathbb{C})$ such that $\|\phi_n(ab) - \phi_n(a)\phi_n(b)\|_2 \to 0$ and $\mathrm{tr}_{k(n)} \circ \phi_n \to \tau$ weak-$*$ (resp. $\|\mathrm{tr}_{k(n)} \circ \phi_n - \tau\|_{A^*} \to 0$).*
(2) *$\tilde{\tau}$ is quasidiagonal (resp. uniform quasidiagonal) if and only if there exist c.c.p. maps $\phi_n : A \to M_{k(n)}(\mathbb{C})$ such that $\|\phi_n(ab) - \phi_n(a)\phi_n(b)\| \to 0$ and $\mathrm{tr}_{k(n)} \circ \phi_n \to \tau$ weak-$*$ (resp. $\|\mathrm{tr}_{k(n)} \circ \phi_n - \tau\|_{A^*} \to 0$).*
(3) *$\tilde{\tau}$ is locally finite dimensional (resp. uniform locally finite dimensional) if and only if there exist c.c.p. maps $\phi_n : A \to M_{k(n)}(\mathbb{C})$ such that $d(a, A_\phi) \to 0$ and $\mathrm{tr}_{k(n)} \circ \phi_n \to \tau$ weak-$*$ (resp. $\|\mathrm{tr}_{k(n)} \circ \phi_n - \tau\|_{A^*} \to 0$).*

PROOF. The key technical fact we will need is again due to Choi-Effros (cf. [**21**, Lemma 3.9]): If $\phi : A \to M_n(\mathbb{C})$ is c.c.p. then the unital linear map $\tilde{\phi} : \tilde{A} \to M_n(\mathbb{C})$ defined by $\tilde{\phi}(a + \lambda) = \phi(a) + \lambda 1$ is also completely positive. With this result in hand all six of the 'if' statements above are simple exercises (left to the reader).

The first four 'only if' statements are completely trivial – just restrict the finite dimensional approximating maps on $\tilde{A}$ to A. The locally finite dimensional case is only slightly harder. One must verify that the multiplicative domain of a map defined on $\tilde{A}$ actually intersects A. For general subalgebras there is no reason that this should be true but luckily A is an ideal in $\tilde{A}$ and hence a routine exercise shows that if $\mathfrak{F} \subset A$ is a given finite set and $B \subset \tilde{A}$ is a subalgebra which nearly contains $\mathfrak{F}$ then B must intersect A and this intersection must almost contain $\mathfrak{F}$ as well. Hence the final two 'only if' statements also follow by simply restricting the maps which are given on $\tilde{A}$. □

Now that we know how to define amenable traces on nonunital C*-algebras we can prove a couple simple facts which will be needed later.

PROPOSITION 3.5.11 (Restriction to an Ideal). *Let $I \triangleleft A$ be an ideal, $\tau \in \mathrm{T}(A)$ and $\gamma = \frac{1}{\|\tau|_I\|}\tau \in \mathrm{T}(I)$. If τ is amenable (resp. uniform amenable) then γ is also amenable (resp. uniform amenable).*

PROOF. If the restriction of τ to I happens to be a state on I then there is nothing to prove since we can just restrict the finite dimensional approximating maps on A to I as in the last proposition. However, that need not be the case so a bit more care is required.

By uniqueness of GNS representations we may identify $\pi_\gamma(I)''$ with the weak closure of $\pi_\tau(I)$ inside $\pi_\tau(A)''$.[19] In the uniform amenable case the proof then follows from the fifth statement in Theorem 3.2.2 since hyperfiniteness passes to all von Neumann subalgebras of a finite, hyperfinite von Neumann algebra.

For the case of amenable traces one applies condition (5) from Theorem 3.1.6 as follows. Let $A \subset B(H)$ be any faithful representation and $\Phi : A \to \pi_\tau(A)''$ be a u.c.p. map such that $\Phi(a) = \pi_\tau(a)$. Now decompose $\pi_\tau(A)'' \cong \pi_\gamma(I)'' \oplus M$ and let $P \in \pi_\tau(A)''$ be the (central) projection such that $P\pi_\tau(A)'' = \pi_\gamma(I)''$. Finally one defines the desired u.c.p. map $\Psi : B(H) \to \pi_\gamma(I)''$ by $\Psi(T) = P\Phi(T)$ and running back through the identifications one sees that $\Psi(x) = \pi_\gamma(x)$ for all $x \in I$. □

PROPOSITION 3.5.12 (Extending from an Ideal). *If $I \triangleleft A$ is an ideal and $\tau \in \mathrm{T}(I)$ is an amenable (resp. uniform amenable) trace then there is a (unique) trace $\gamma \in \mathrm{T}(A)$ which extends τ and it is amenable (resp. uniform amenable).*

PROOF. Since I is an ideal there is a unique extension of the GNS representation $\pi_\tau : I \to B(L^2(I,\tau))$ to a *-homomorphism $\sigma : A \to \pi_\tau(I)''$. Let γ be the unique tracial state extension of τ to A (which is gotten by composing the GNS vector state on $\pi_\tau(I)''$ with the map σ). Since σ is unitarily equivalent to $\pi_\gamma : A \to B(L^2(A,\gamma))$ it follows that γ is uniform amenable whenever τ is since hyperfiniteness is obviously passed to $\pi_\gamma(A)''$.

Now assume that τ is only an amenable trace and fix a faithful *-representation $A \subset B(H)$. To see the essence of the proof we will also further assume that I is an essential ideal in A and hence belongs to the strict closure of I. Now if $\Phi : B(H) \to \pi_\tau(I)''$ is a u.c.p. map such that $\Phi(x) = \pi_\tau(x)$ for all $x \in I$ then it suffices to show that $\Phi(a) = \sigma(a)$ for all $a \in A$. But for each $a \in A$ we can find a sequence $\{x_n\} \subset I$ such that $yx_n \to ya$ and $x_n y \to ay$ (in norm) for all $y \in I$. Since I falls in the multiplicative domain of Φ it follows that $\pi_\tau(y)\pi_\tau(x_n) \to \pi_\tau(y)\Phi(a)$

---

[19]Actually, since I is an ideal we get the decomposition $A^{**} = I^{**} \oplus (A/I)^{**}$ and hence we can identify $\pi_\gamma(I)''$ with a direct summand of $\pi_\tau(A)''$.

and $\pi_\tau(x_n)\pi_\tau(y) \to \Phi(a)\pi_\tau(y)$ for all $y \in I$ and hence $\Phi(a)$ is the strict limit of $\pi_\tau(x_n)$. In other words, Φ is automatically strictly continuous and hence must agree with σ (since σ is also strictly continuous and agrees with Φ on I). This shows that γ is an amenable trace on A as well.

The argument in the previous paragraph actually shows that amenable traces always extend to amenable traces on the multiplier algebra of a nonunital C*-algebra. Hence when $I \triangleleft A$ is not an essential ideal we can uniquely extend an amenable trace on I to one on A by first extending to the multiplier algebra of I, denoted $M(I)$, and then composing with the canonical *-homomorphism $A \to M(I)$. □

CHAPTER 4

Examples and special cases

Having defined so many subsets of traces it may be worthwhile to consider a number of examples. Our hope is that a detailed look at these examples will not only help cement the definitions in the reader's mind but also begin to illustrate how these traces are related to various other problems and conjectures.

4.1. Discrete groups

As usual, groups provide some interesting and instructive examples. Since we are sticking to unital algebras, we will only consider the discrete case.

Following standard notation, we will let $C^*(\Gamma)$ and $C_r^*(\Gamma)$ denote the full (i.e. universal) and reduced C*-algebra, respectively, of a discrete group Γ. When studying amenable traces it turns out that there is an enormous difference between considering the full and reduced algebras. For $C_r^*(\Gamma)$ there is a sort of 'all or nothing' principle which asserts that either every trace on $C_r^*(\Gamma)$ is amenable or no trace on $C_r^*(\Gamma)$ is amenable – the first case occurring if and only if Γ is an amenable group.

PROPOSITION 4.1.1. (Reduced group C*-algebras – compare with [**6**, Corollary 2.11].) *For a discrete group Γ the following are equivalent:*

(1) Γ *is amenable.*
(2) $\mathrm{T}(C_r^*(\Gamma)) = \mathrm{AT}(C_r^*(\Gamma))$.
(3) $\mathrm{AT}(C_r^*(\Gamma)) \neq \emptyset$.

PROOF. (1) $\Longrightarrow$ (2). A result of Lance [**49**] states that $C_r^*(\Gamma)$ is nuclear whenever Γ is amenable. Hence for every C*-algebra B there is a unique C*-norm on the algebraic tensor product $C_r^*(\Gamma) \odot B$. In particular, $C_r^*(\Gamma) \otimes B$ enjoys the universal property of the maximal tensor product[1] and hence it is clear from condition (4) of Theorem 3.1.6 that every trace on $C_r^*(\Gamma)$ is amenable.

(2) $\Longrightarrow$ (3) is immediate since discreteness implies that $C_r^*(\Gamma)$ always has a tracial state.

(3) $\Longrightarrow$ (1). Suppose that τ is an amenable trace on $C_r^*(\Gamma)$. Then we may extend τ to a state ϕ on $B(l^2(\Gamma))$ such that

$$\phi(\lambda_g T \lambda_g^*) = \phi(T)$$

for all $g \in \Gamma$ and $T \in B(l^2(\Gamma))$. A well known calculation shows that if we regard $l^\infty(\Gamma) \subset B(l^2(\Gamma))$ as multiplication operators then the left translation action of Γ on $l^\infty(\Gamma)$ is just given by conjugation: $f \mapsto \lambda_g f \lambda_g^*$. Hence the restriction of the state ϕ to $l^\infty(\Gamma)$ defines a classical invariant mean which implies Γ is amenable. $\square$

---

[1]Every pair of representations $\pi : C_r^*(\Gamma) \to B(H)$ and $\sigma : B \to B(H)$ with commuting ranges induces a natural *-homomorphism $C_r^*(\Gamma) \otimes B \to B(H)$.

COROLLARY 4.1.2. Γ *is not amenable if and only if $C_r^*(\Gamma)$ has no amenable traces.*

Jonathan Rosenberg first observed that if $C_r^*(\Gamma)$ is quasidiagonal then Γ must be amenable. He has conjectured that the converse holds as well. While this conjecture remains elusive we now point out that (a) Rosenberg's observation is easily deduced from amenable trace considerations and (b) his conjecture would follow from a similar statement about traces.

First suppose that $C_r^*(\Gamma)$ is quasidiagonal. Then $C_r^*(\Gamma)$ would have to have at least one amenable trace (cf. Proposition 3.3.2) and hence, by the last proposition, Γ would be amenable. Note that this argument requires much less than quasidiagonality since this was only used to get the existence of an amenable trace.

Now to the tracial statement that would imply Rosenberg's conjecture.

PROPOSITION 4.1.3. (Rosenberg's Conjecture) *If the canonical trace on $C_r^*(\Gamma)$ is quasidiagonal then $C_r^*(\Gamma)$ is quasidiagonal. In particular, if this is the case for every amenable group then Rosenberg's conjecture would follow.*

PROOF. This is actually a special case of a more general result: If $\tau \in \mathrm{AT}(A)_{\mathrm{QD}}$ is faithful then A is quasidiagonal. Indeed, assume that $\phi_n : A \to M_{k(n)}(\mathbb{C})$ are u.c.p. maps which are asymptotically multiplicative (in norm) and recover τ. Since these maps are almost multiplicative in norm we get a τ-preserving $*$-homomorphism into the quotient of the norm bounded sequences by the ideal of sequences which tend to zero in norm

$$\frac{\Pi M_{k(n)}(\mathbb{C})}{\oplus M_{k(n)}(\mathbb{C})}$$

similar to the proof of (1) $\Longrightarrow$ (2) from Theorem 3.1.7. Since τ is faithful so is the induced $*$-homomorphism

$$A \to \frac{\Pi M_{k(n)}(\mathbb{C})}{\oplus M_{k(n)}(\mathbb{C})}$$

and hence $\|a\| = \limsup \|\phi_n(a)\|$ for every $a \in A$. From Voiculescu's abstract characterization of quasidiagonality it follows that A must be quasidiagonal.[2] □

Now we consider the full group C*-algebras $C^*(\Gamma)$ where things are totally different. For example, this C*-algebra always has at least one amenable trace. In fact, the trivial representation can be regarded as a (uniform locally) finite dimensional trace on $C^*(\Gamma)$.

There is another trace on $C^*(\Gamma)$ coming from the left regular representation and, in contrast to the reduced C*-algebra case, this trace is also amenable for a large class of non-amenable discrete groups. For the remainder of this section we will let τ denote the trace on $C^*(\Gamma)$ coming from the left regular representation.

PROPOSITION 4.1.4. (Residually Finite Groups) *If Γ is residually finite then τ is a locally finite dimensional trace on $C^*(\Gamma)$. In particular, it is amenable.*

---

[2] The careful reader may be worried about only having $\|a\| = \limsup \|\phi_n(a)\|$ instead of an honest limit. However it is a routine exercise to replace the ϕ_n's with asymptotically multiplicative maps satisfying the right limit condition by passing to subsequences and taking direct sums.

PROOF. We will show that there are finite dimensional $*$-representations $\pi_n : C^*(\Gamma) \to M_{k(n)}(\mathbb{C})$ such that for each non-trivial group element $g \in \Gamma$ we have

$$\operatorname{tr}_{k(n)} \circ \pi_n(\lambda_g) \to 0.$$

Since the linear span of such unitaries, together with the unit, is dense in $C^*(\Gamma)$ it will follow that τ is the weak-$*$ limit of finite dimensional traces and the proof will be complete.

To construct such representations we let $\Gamma_1 \trianglerighteq \Gamma_2 \trianglerighteq \ldots$ be a descending sequence of normal subgroups each of which has finite index in Γ and such that their intersection is the neutral element. Let $\pi_n : C^*(\Gamma) \to B(l^2(\Gamma/\Gamma_n))$ be the unitary representation induced by the left regular representation of Γ/Γ_n. Since Γ/Γ_n is a finite group, $B(l^2(\Gamma/\Gamma_n))$ is finite dimensional and a standard calculation completes the proof. □

COROLLARY 4.1.5. (Maximally Almost Periodic Groups) *If Γ is maximally almost periodic[3] then τ is a locally finite dimensional trace on $C^*(\Gamma)$.*

PROOF. It is well known that every discrete maximally almost periodic group is the increasing union of its residually finite subgroups.[4] If $\Lambda \subset \Gamma$ is a subgroup then $C^*(\Lambda) \subset C^*(\Gamma)$ and the restriction of the canonical trace on $C^*(\Gamma)$ is just the canonical trace on $C^*(\Lambda)$. With these observations the remainder of the proof amounts to a standard approximation argument. Indeed, if $\mathfrak{F} \subset C^*(\Gamma)$ is a finite set then, after a small perturbation, we may assume that $\mathfrak{F} \subset C^*(\Lambda) \subset C^*(\Gamma)$ where Λ is a residually finite subgroup of Γ. By the previous corollary we can find maps from $C^*(\Lambda)$ to matrices whose multiplicative domains almost contain $\mathfrak{F}$ and recover the canonical trace. We then extend these maps to all of $C^*(\Gamma)$ by Arveson's extension theorem. □

We are now in a position to illustrate the difference between being an amenable trace and a uniform amenable trace.

PROPOSITION 4.1.6. *If τ is a uniform amenable trace on $C^*(\Gamma)$ then Γ is amenable.*

PROOF. If τ is a uniform amenable trace then statement (2) of Proposition 3.5.8 implies that the *reduced* group C*-algebra also has a (uniform) amenable trace and hence Γ is amenable. □

REMARK 4.1.7. Propositions 4.1.4 and 4.1.6 imply that the sets UAT(A), UAT(A)$_{\mathrm{QD}}$ and UAT(A)$_{\mathrm{LFD}}$ need not be weak-$*$ closed in general. Indeed, if Γ is residually finite and non-amenable then τ arises as the weak-$*$ limit of (uniform locally) finite dimensional traces however it is *not* uniform amenable.

Proposition 4.1.4 and Corollary 4.1.5 are well known (cf. [**47**], [**81**]) but usually formulated in terms of *Kirchberg's factorization property*.

---

[3]In other words, Γ is isomorphic to a subgroup of a compact group. This class of groups contains all residually finite groups.

[4]Since all finitely generated linear groups are residually finite (cf. [**1**]) and maximally almost periodic groups have a separating family of homomorphisms into linear groups it can be shown that every finitely generated maximally almost periodic group is residually finite. Since every discrete group is the union of its finitely generated subgroups the claim follows.

DEFINITION 4.1.8. A discrete group Γ has *Kirchberg's factorization property* if the natural $*$-representation

$$C^*(\Gamma) \odot C^*(\Gamma) \to B(l^2(\Gamma))$$

induced by the left and right regular representations is continuous with respect to the minimal tensor product norm.[5]

The following theorem is an immediate consequence of Theorem 3.1.6. Indeed, the 'right regular' representation used in that theorem is nothing but the obvious generalization of the right regular representation of a group.

THEOREM 4.1.9. (Kirchberg) Γ *has the factorization property if and only if* τ *is an amenable trace on* $C^*(\Gamma)$.

Evidently all amenable groups have the factorization property (since $C^*(\Gamma) = C_r^*(\Gamma)$ is nuclear) but there are plenty of non-amenable examples as well. For example, all residually finite groups (e.g. free groups, $SL(n,\mathbb{Z})$, etc.) and, more generally, maximally almost periodic groups. With this observation in hand we can now recover a result of Bekka which states that Rosenberg's conjecture has an affirmative answer in the case that Γ is amenable and maximally almost periodic. (See [25, Corollary 4] for another proof of this result.)

COROLLARY 4.1.10. (Bekka) *If* Γ *is amenable and maximally almost periodic then* $C^*(\Gamma) = C_r^*(\Gamma)$ *is quasidiagonal.*

PROOF. Since Γ is amenable the full and reduced C*-algebras are equal. Hence Proposition 4.1.5 implies the canonical trace on $C_r^*(\Gamma)$ is quasidiagonal and so we can apply Proposition 4.1.3. □

The main result of [47] asserts that every discrete group which has both the factorization property and Kazhdan's property T must be residually finite. The main technical tool in the proof is the following fact (see [47, Proposition 2.3] for a proof[6]).

LEMMA 4.1.11. (Kirchberg) *If* Γ *has property T and* $\phi_n : C^*(\Gamma) \to M_{k(n)}(\mathbb{C})$ *are u.c.p. maps such that* $\|\phi_n(ab) - \phi_n(a)\phi_n(b)\|_2 \to 0$ *for all* $a, b \in C^*(\Gamma)$ *then there exist* $*$-*homomorphisms* $\pi_n : C^*(\Gamma) \to M_{l(n)}(\mathbb{C})$ *such that*

$$|\mathrm{tr}_{k(n)}(\phi_n(a)) - \mathrm{tr}_{l(n)}(\pi_n(a))| \to 0$$

for all $a \in C^*(\Gamma)$.

Given this lemma it is a simple matter to provide the proof of the following fact.

PROPOSITION 4.1.12. *If* Γ *is a property T group then*

$$\mathrm{AT}(C^*(\Gamma)) = \mathrm{AT}(C^*(\Gamma))_{\mathrm{QD}} = \mathrm{AT}(C^*(\Gamma))_{\mathrm{LFD}}.$$

---

[5]In other words, the $*$-representation defined on the maximal tensor product factors through the minimal tensor product.

[6]The reader acquainted with Property T arguments may find this a fun exercise. Basically one applies Stinespring to a u.c.p. map $\phi : A \to M_n(\mathbb{C})$. If ϕ is almost multiplicative in 2-norm then the Stinespring projection gives a Hilbert-Schmidt operator which is almost invariant under the conjugation action. One perturbs, by Property T, to a Hilbert-Schmidt operator in the commutant of the Stinespring representation and some functional calculus and other standard estimates complete the proof.

Since the factorization property is equivalent to knowing that τ is an amenable trace on $C^*(\Gamma)$ we can combine Kirchberg's result quoted above with some work of Olshanskii [56] to get the following connection between amenable traces and one of the major open problems in geometric group theory.

PROPOSITION 4.1.13. *Every hyperbolic group is residually finite if and only if τ is an amenable trace on $C^*(\Gamma)$ for every hyperbolic group Γ.*

PROOF. Since the canonical trace is always amenable for residually finite groups we only need to observe the 'if' statement. However in [56] Olshanskii shows, among other things, that every hyperbolic group can be embed into a hyperbolic group with property T. Since, for property T groups, residual finiteness and the factorization property are equivalent the proposition follows. □

Finally we consider the case of free groups. As observed in [45] understanding the space of amenable traces on $C^*(\mathbb{F}_n)$ is of paramount importance.[7] For now we just observe that, like the case of property T groups, some of the sets of traces defined here coincide in this case.

PROPOSITION 4.1.14. *For a free group we have*
$$\mathrm{AT}(C^*(\mathbb{F}_n)) = \mathrm{AT}(C^*(\mathbb{F}_n))_{\mathrm{QD}} = \mathrm{AT}(C^*(\mathbb{F}_n))_{\mathrm{LFD}}.$$

PROOF. This follows from a simple observation of Kirchberg (see [45, Lemma 4.5] or [39]). Namely, since R^ω is a von Neumann algebra every unitary can be lifted to a unitary in $l^\infty(R)$. A standard approximation argument yields the following improvement: If $\pi : C^*(\mathbb{F}_n) \to R^\omega$ is a $*$-homomorphism then there exist $*$-homomorphisms $\pi_n : C^*(\mathbb{F}_n) \to M_{k(n)}(\mathbb{C})$ such that
$$|\tau_\omega \circ \pi(a) - \mathrm{tr}_{k(n)} \circ \pi_n(a)| \to 0$$
for all $a \in C^*(\mathbb{F}_n)$.[8] This evidently implies that every amenable trace arises as the weak-$*$ limit of finite dimensional traces and the proof is complete. □

4.2. Nuclear and WEP C*-algebras

We now see how the theory of amenable traces develops for several important classes of examples. We begin with two where it turns out that every trace enjoys some sort of amenability – nuclear C*-algebras and C*-algebras with the weak expectation property (WEP) of Lance (cf. [49]).

In the proof of Proposition 4.1.1 we saw that every trace on a nuclear C*-algebra is amenable. This result is a trivial consequence of the original tensor product definition of nuclearity[9] and Lance's tensor product trick from the proof of Theorem 3.1.6. A far more substantial result, however, is the fact that every trace on a nuclear C*-algebra is *uniform* amenable.

THEOREM 4.2.1. (Nuclear C*-algebras) *If A is nuclear then*
$$\mathrm{T}(A) = \mathrm{AT}(A) = \mathrm{UAT}(A).$$

---

[7]The main question, which turns out to be equivalent to Connes' embedding problem, is whether or not every trace is amenable on $C^*(\mathbb{F}_\infty)$.

[8]First lift the generating unitaries to unitaries in $l^\infty(R)$ and then approximate them (in 2-norm) by unitaries in large finite dimensional subfactors of R.

[9]A is nuclear if and only if there is a unique C*-norm on $A \odot B$ for every C*-algebra B.

PROOF. This is an immediate consequence of Theorem 3.2.2 and the celebrated fact that an injective von Neumann algebra is hyperfinite (cf. [**23**], [**33**], [**65**]) since it is not hard to show that every representation of a nuclear C*-algebra yields an injective von Neumann algebra (use Lance's trick from the proof of Theorem 3.1.6 to show the commutant is injective). □

Recall that a C*-algebra A is said to have the *weak expectation property* (WEP) if for every *faithful* representation $\pi : A \to B(H)$ there exists a u.c.p. map $\Phi : B(H) \to \pi(A)''$ such that $\Phi(\pi(a)) = \pi(a)$, for all $a \in A$. This is a large class of C*-algebras (including every injective C*-algebra and every nuclear C*-algebra). In fact, [**45**, Corollary 3.5] states that every (separable) C*-algebra is contained in a (separable) simple C*-algebra with the WEP (in particular, the WEP class contains lots of separable non-nuclear C*-algebras).

PROPOSITION 4.2.2. (C*-algebras with the WEP) *If A has the WEP then* $T(A) = AT(A)$.

PROOF. Let $\tau \in T(A)$ be given and assume that $A \subset A^{**} \subset B(H)$ is the universal representation.[10] Applying the definition of the weak expectation property to this representation we get a u.c.p. map $\Phi : B(H) \to A^{**}$ such that $\Phi(a) = a$ for all $a \in A$. In particular, A falls in the multiplicative domain of Φ and hence, letting τ^{**} denote the normal extension of τ to A^{**}, we have that $\tau^{**} \circ \Phi$ is a state on $B(H)$ which extends τ and contains A in its centralizer. □

Though the previous proposition is a simple consequence of the definitions it turns out to have some interesting consequences. Indeed, a common phenomenon in mathematics is that certain problems can be easily answered once the right point of view is found. Amenable traces turn out to be the "right" concept to consider for a number of operator algebraic questions as we will see in this paper.

Our first application is related to Kirchberg's remarkable embedding theorem: A C*-algebra is exact if and only if it embeds into the Cuntz algebra $\mathcal{O}_2$. It is natural to wonder if A is exact and has additional properties whether or not A can be embed into a nuclear C*-algebra with the same additional properties. For example, it is not known whether or not every exact, quasidiagonal C*-algebra can be embed into a nuclear, quasidiagonal C*-algebra. (See, however, [**57**] where cones and suspensions over exact C*-algebras are shown to be embeddable into AF algebras!) On the other hand, some experts have already observed that the stably finite, exact C*-algebra $C_r^*(\mathbb{F}_2)$ can't be embed into a stably finite nuclear C*-algebra.[11] The proof typically quoted depends on the uniqueness of the trace on $C_r^*(\mathbb{F}_2)$, Haagerup's remarkable result that every unital, stably finite, exact C*-algebra has a tracial state ([**34**], [**36**]) together with Connes' theorem that injective implies hyperfinite and hence is highly non-trivial. On the other hand, amenable

---

[10] This violates our standing separability assumption, but otherwise causes no harm.

[11] Whether or not every exact, stably finite C*-algebra can be embed into a nuclear, stably finite C*-algebra was a problem pondered by some experts a few years ago. This problem naturally arose from an important question of Blackadar-Kirchberg which asks whether or not every nuclear, stably finite C*-algebra is quasidiagonal. A natural way to provide counterexamples to the quasidiagonal question is to try to embed a (necessarily exact and stably finite) non-quasidiagonal C*-algebra into a stably finite, nuclear C*-algebra. Unfortunately this strategy is doomed to failure for all of the examples of exact, stably finite, non-quasidiagonal C*-algebras that we currently know – see Proposition 6.6.1.

traces show that this is a general fact about non-amenable discrete groups. (See also Bedøs' observations along these lines [**6**, Corollary 2.11].)

COROLLARY 4.2.3. *Let Γ be a discrete group. One can find a $*$-homomorphism $\pi : C_r^*(\Gamma) \to A$, where A has the WEP and at least one tracial state, if and only if Γ is amenable.*

PROOF. If Γ is amenable then $C_r^*(\Gamma)$ is nuclear (hence WEP) and has a trace. On the other hand, since every trace on A is amenable, the existence of π would imply that $C_r^*(\Gamma)$ has an amenable trace (just restrict the trace on A to $\pi(C_r^*(\Gamma))$) and hence, by Proposition 4.1.1, Γ would be amenable. □

COROLLARY 4.2.4. *If Γ is a discrete, non-amenable group then $C_r^*(\Gamma)$ can't be embed into a stably finite, nuclear C^*-algebra or into a finite, hyperfinite von Neumann algebra.*

PROOF. Hyperfinite, finite von Neumann algebras are injective (hence WEP) and have a tracial state. Every unital, stably finite, nuclear C*-algebra has a tracial state (cf. [**36**]). □

Note, of course, that not only are embeddings impossible but one can't even find a non-zero $*$-homomorphism from $C_r^*(\Gamma)$ to a stably finite, nuclear C*-algebra when Γ is non-amenable. This result is in contrast to the following theorem of Kirchberg: Every exact C*-algebra admits a complete order embedding into the CAR algebra (cf. [**81**, Theorem 9.1]). All together these results say that if Γ is exact and non-amenable then $*$-homomorphic embeddings of $C_r^*(\Gamma)$ into purely infinite, nuclear algebras always exist; complete order embeddings into stably finite, nuclear algebras always exist; but, non-zero $*$-homomorphisms to stably finite, nuclear algebras *never* exist.

We now use amenable traces to give a simple proof of a generalization of [**8**, Theorem 6.8]. If $\pi : G \to B(H)$ is a unitary representation of a locally compact group G then Bekka defines the *representation* to be amenable if there exists a state φ on $B(H)$ such that $\varphi(\pi(g)T\pi(g^{-1})) = \varphi(T)$ for all $T \in B(H)$ and for all $g \in G$. In the language of this paper π is amenable if and only if $A = C^*(\{\pi(g) : g \in G\}) = span\{\pi(g) : g \in G\}^-$ (norm closure, since G is a group) has an amenable trace (it is a simple exercise to show that this is equivalent to Bekka's definition). In [**8**, Theorem 6.8] Bekka proves the following result in the case that A is nuclear.

COROLLARY 4.2.5. *Let $\pi : G \to B(H)$ be a strongly continuous, unitary representation of a locally compact group G and $A = C^*(\{\pi(g) : g \in G\})$. If A has the WEP then π is an amenable representation if and only if A has a tracial state.*

PROOF. Strictly speaking one might be concerned about separability issues here, but we only assume separability for convenience and this is not necessary for virtually everything discussed in this paper. If π is amenable then, by definition, A has an amenable trace. If A has the WEP and a tracial state then this trace is amenable and hence π is amenable. □

REMARK 4.2.6. In the paragraph preceding [**8**, Theorem 6.8], Bekka asked if there is any general relationship between the amenability of π and the nuclearity of $A = C^*(\{\pi(g) : g \in G\})$. The answer is that – in general – there is absolutely no relationship. That is, it is possible that A be nuclear while π is not amenable (let A be a Cuntz algebra, for example, and Γ any dense subgroup of the unitary group of A) and that A be non-nuclear while π is amenable (take A to be any non-nuclear Popa algebra and then A must have at least one quasidiagonal trace since it is a quasidiagonal C*-algebra).

4.3. Locally reflexive, exact and quasidiagonal C*-algebras

We regard amenable traces as corresponding to nice GNS representations ('nice' is quite ambiguous at the moment). One natural question to ask is whether or not being 'nice' can be characterized in terms of the von Neumann algebra one gets in the GNS representation. Unfortunately this is not the case, in general, since the canonical traces on $C^*(\mathbb{F}_2)$ and $C^*_r(\mathbb{F}_2)$ yield the same free group factor while the trace on $C^*(\mathbb{F}_2)$ is amenable (by residual finiteness) and the trace on $C^*_r(\mathbb{F}_2)$ is not amenable (by non-amenability of free groups). Thus it is the *representation* (including the domain), as opposed to the *image* of the representation, which matters in general. However, if one assumes some additional structure on the algebra then the *image* of the representation does in fact characterize 'niceness'. We will see another example of this later on (cf. Proposition 6.3.4): If A has the so-called local lifting property then a trace on A is amenable if and only if the corresponding GNS von Neumann algebra admits a trace preserving embedding into the ultraproduct of the hyperfinite II$_1$-factor. In this section we show that assuming local reflexivity gives a much sharper statement (see Corollary 4.3.4).

The notion of local reflexivity in the context of C*-algebras is due to Effros and Haagerup [**28**]. A remarkable result of Kirchberg asserts that every exact C*-algebra has this property [**46**] and he has conjectured that the converse holds as well.

DEFINITION 4.3.1. *A is* locally reflexive *if for each finite dimensional operator system $X \subset A^{**}$ there exists a net of u.c.p. maps $\phi_\lambda : X \to A$ such that*

$$\eta(\phi_\lambda(x)) \to \eta(x)$$

for all $x \in X$ and $\eta \in A^*$. (i.e. $\phi_\lambda \to id_X$ in the point weak- topology.)*

The following improvement is a standard Hahn-Banach type argument but we will soon find it quite useful.

LEMMA 4.3.2. *Assume A is locally reflexive and $\mathfrak{F} \subset A$ is a finite set. Then for any finite dimensional operator system $\mathfrak{F} \subset X \subset A^{**}$ there exists a net of u.c.p. maps $\phi_\lambda : X \to A$ such that*

$$\eta(\phi_\lambda(x)) \to \eta(x)$$

for all $x \in X$ and $\eta \in A^*$ and such that $\|a - \phi_\lambda(a)\| \to 0$ for each $a \in \mathfrak{F}$.

PROOF. Assume first that $\mathfrak{F}$ consists of a single element $a \in A$. Applying the Hahn-Banach theorem to the convex hull of the u.c.p. maps provided by the definition of locally reflexivity one can find a net converging in norm. For general finite sets one employs the usual trick of taking direct sums and reducing to the singleton case (as we already saw in the proof of (1) $\implies$ (2) from Theorem 3.1.6, for example). □

The main result concerning amenable traces on locally reflexive C*-algebras is as follows.

THEOREM 4.3.3. *Assume that A is locally reflexive. Then it is always the case that* $\mathrm{AT}(A) = \mathrm{UAT}(A)$ *and* $\mathrm{AT}(A)_{\mathrm{QD}} = \mathrm{UAT}(A)_{\mathrm{QD}}$.[12]

PROOF. We only give the proof of $\mathrm{AT}(A)_{\mathrm{QD}} = \mathrm{UAT}(A)_{\mathrm{QD}}$ as it will be clear that essentially the same proof gives the other equality.

So let $\tau \in \mathrm{AT}(A)_{\mathrm{QD}}$ be arbitrary. Evidently it suffices to prove that if $\mathfrak{F} \subset A$ is an arbitrary finite set and $\varepsilon > 0$ then there exists a u.c.p. map $\varphi : A \to B$, where B is a finite dimensional C*-algebra, such that $\|\varphi(xy) - \varphi(x)\varphi(y)\| < \varepsilon$, for all $x, y \in \mathfrak{F}$ and such that there exists a trace γ on B such that $\|\tau - \gamma \circ \varphi\|_{A^*} < \varepsilon$ (cf. Lemma 2.5).

In order to do this, we will show that for each finite dimensional operator system $X \subset A^{**}$ containing both the set $\mathfrak{F}$ and $\{ab : a, b \in \mathfrak{F}\}$, there exists a sequence of normal, u.c.p. maps $\psi_n : A^{**} \to M_{s(n)}(\mathbb{C})$ such that $\mathrm{tr}_{s(n)} \circ \psi_n(x) \to \tau^{**}(x)$, for all $x \in X$ and *each ψ_n is ε-multiplicative on $\mathfrak{F}$*. If we are able to do this then one can construct a net of normal, u.c.p. maps $\varphi_\lambda : A^{**} \to M_{k(\lambda)}(\mathbb{C})$ with the property that $\mathrm{tr}_{k(\lambda)} \circ \varphi_\lambda \in (A^{**})_* = A^*$ for all λ, each φ_λ is ε-multiplicative on $\mathfrak{F}$ and (here is the key) $\mathrm{tr}_{k(\lambda)} \circ \varphi_\lambda \to \tau^{**}$ in the weak topology coming from A^{**}. Hence, by the Hahn-Banach theorem, τ^{**} belongs to the *norm* closure of the convex hull of $\{\mathrm{tr}_{k(\lambda)} \circ \varphi_\lambda\} \subset A^*$. Then one would be able to choose a finite set $\lambda_1, \ldots, \lambda_p$ and positive real numbers $\theta_1, \ldots, \theta_p$ such that $\sum \theta_i = 1$ and $\|\tau^{**} - \sum \theta_i \mathrm{tr}_{k(\lambda_i)} \circ \varphi_{\lambda_i}\|_{A^*} < \varepsilon$. Finally one would define $B = M_{k(\lambda_1)}(\mathbb{C}) \oplus \cdots \oplus M_{k(\lambda_p)}(\mathbb{C})$, $\varphi = \varphi_{\lambda_1} \oplus \cdots \oplus \varphi_{\lambda_p}$ and $\gamma = \sum \theta_i \mathrm{tr}_{k(\lambda_i)}$. Since we arranged that φ_λ is ε-multiplicative on $\mathfrak{F}$ *for every* λ, it is clear that φ will also be close to multiplicative on $\mathfrak{F}$.

So let $X \subset A^{**}$ be any finite dimensional operator system containing the sets $\mathfrak{F}$ and $\{ab : a, b \in \mathfrak{F}\}$. Since $\tau \in \mathrm{AT}(A)_{\mathrm{QD}}$ we can find a sequence of u.c.p. maps $\varphi_m : A \to M_{k(m)}(\mathbb{C})$ which are asymptotically multiplicative (in norm) and which recapture τ (as a weak-$*$ limit) after composing with the traces on $M_{k(m)}(\mathbb{C})$. Note that by passing to a subsequence, if necessary, we may further assume that each φ_m is as close to multiplicative as one likes on the set $\mathfrak{F}$. Since A is locally reflexive, we can find a net of u.c.p. maps $\beta_t : X \to A$ such that $\beta_t(x) \to x$ in the weak-$*$ topology (coming from A^*) for all $x \in X$. By the lemma preceding this theorem we may further assume that $\|a - \beta_t(a)\| \to 0$ for all $a \in \mathfrak{F} \cup \{ab : a, b \in \mathfrak{F}\}$. Passing to a subnet, if necessary, we may also assume that $\|\beta_t(a) - a\| < \varepsilon$ for all $a \in \mathfrak{F} \cup \{ab : a, b \in \mathfrak{F}\}$ and for all t. In particular, this implies that $\varphi_m \circ \beta_t$ is nearly multiplicative on $\mathfrak{F}$ for all t, m.

We almost have the desired maps ψ_n. Since X is finite dimensional we can choose a linear basis $\{x_1, \ldots, x_q\}$. For each $n \in \mathbb{N}$ first choose $t(n)$ such that $|\tau^{**}(x_i) - \tau(\beta_{t(n)}(x_i))| < 1/n$ for $1 \leq i \leq q$. Next, choose m_n such that

$$|\tau(\beta_{t(n)}(x_i)) - \mathrm{tr}_{k(m_n)} \circ \varphi_{m_n}(\beta_{t(n)}(x_i))| < 1/n$$

for $1 \leq i \leq q$. Then defining $\tilde{\psi}_n = \varphi_{m_n} \circ \beta_{t(n)} : X \to M_{k(m_n)}(\mathbb{C})$ we have that $\mathrm{tr}_{k(m_n)} \circ \tilde{\psi}_n(x) \to \tau^{**}(x)$ for all $x \in X$.

By Arveson's extension theorem, we may assume that each $\tilde{\psi}_n$ is actually defined on all of A^{**}. The only problem is that we can't be sure that the Arveson extensions are normal on A^{**}. However a tiny perturbation of the $\tilde{\psi}_n$ will yield

---

[12] As previously mentioned, we don't know whether $\mathrm{AT}(A)_{\mathrm{LFD}} = \mathrm{UAT}(A)_{\mathrm{LFD}}$ for all locally reflexive A – this is a very important open problem (see Section 6.1).

normal maps ψ_n with all the right properties. (Use the fact that c.p. maps to matrix algebras are nothing but positive linear functionals on matrices over the given algebra and that the set of normal linear functionals on a von Neumann algebra is dense in the dual space.) □

Theorem 3.2.2 together with the result above gives our first corollary.

COROLLARY 4.3.4. *Let A be locally reflexive and τ be a tracial state on A. Then τ is amenable if and only if $\pi_\tau(A)''$ is hyperfinite.*

The next corollary was proved in [**45**, Theorem 7.5]. Our proof is not any simpler than Kirchberg's but it emphasizes the role of amenable traces and representation theory. Kirchberg, in fact, also used (classical) invariant means but his emphasis is on exactness and tensor products (cf. [**45**, Proposition 7.1]).

COROLLARY 4.3.5. *Let Γ be a discrete group with the factorization property. Then Γ is amenable if and only if $C^*(\Gamma)$ is locally reflexive. In particular, if Γ is any residually finite, non-amenable group then $C^*(\Gamma)$ is not locally reflexive (hence not exact).*

PROOF. If Γ is amenable then $C^*(\Gamma)$ is nuclear and hence locally reflexive. If $C^*(\Gamma)$ is locally reflexive and τ is an amenable trace then $\pi_\tau(C^*(\Gamma))''$ is a hyperfinite von Neumann algebra. Hence the canonical trace on the *reduced* algebra $C_r^*(\Gamma)$ is also amenable and thus Γ is amenable. □

We won't actually need condition (3) in the next corollary however, it may be of independent interest. Pisier gave the first proof of (3) $\Longrightarrow$ (4), though Ozawa later observed that exactness was an unnecessary assumption (see Theorem 3.2.2).

COROLLARY 4.3.6. *Let A be an exact C^*-algebra and $\tau \in$ T(A). Then the following are equivalent:*
  (1) *$\tau \in$ AT(A).*
  (2) *$\tau \in$ UAT(A).*
  (3) *The GNS representation is a nuclear map into $\pi_\tau(A)''$. That is, there exist u.c.p. maps $\phi_n : A \to M_{k(n)}(\mathbb{C})$ and $\psi_n : M_{k(n)}(\mathbb{C}) \to \pi_\tau(A)''$ such that $\|\pi_\tau(a) - \psi_n \circ \phi_n(a)\| \to 0$ for all $a \in A$.*
  (4) *$\pi_\tau(A)''$ is hyperfinite.*

PROOF. Since exact C*-algebras are locally reflexive (cf. [**46**, pg. 71]) we have the equivalence of (1), (2) and (4). Since nuclearity obviously implies weak nuclearity we also get the implication (3) $\Longrightarrow$ (4) from Theorem 3.2.2. Hence we are only left to prove (1) $\Longrightarrow$ (3).

Let $A \subset B(H)$ be any faithful representation of A. Then the inclusion $A \hookrightarrow B(H)$ is a nuclear map (cf. [**81**, Theorem 7.3]). From part (5) of Theorem 3.1.6 there exists a u.c.p. map $\Phi : B(H) \to \pi_\tau(A)''$ such that $\Phi(a) = \pi_\tau(a)$, for all $a \in A$. It follows that π_τ (which can be identified with $\Phi|_A$ composed with $A \hookrightarrow B(H)$) is a nuclear map. □

It is an important open problem to determine which quasidiagonal C*-algebras satisfy the equation T(A) = AT(A)$_{\text{QD}}$. Indeed, if $A = C^*(\mathbb{F}_\infty)$ then the equation T(A) = AT(A)$_{\text{QD}}$ is equivalent to Connes' embedding problem (see Section 6.3). If A is nuclear and quasidiagonal then this equation is predicted by Elliott's conjecture (see Proposition 6.1.21). When we first began studying amenable traces we thought that it may be possible to prove Connes' problem by verifying the equation T(A) = AT(A)$_{\text{QD}}$ for all quasidiagonal C*-algebras. Support for this strategy is provided by the following observation.

PROPOSITION 4.3.7. *If A is quasidiagonal and $\tau \in$ T(A) then there exists a sequence of u.c.p. maps $\phi_n : A \to M_{k(n)}(\mathbb{C})$, which are asymptotically multiplicative in norm, and a sequence of states $\sigma_n \in S(M_{k(n)}(\mathbb{C}))$ such that $\sigma_n \circ \phi_n(a) \to \tau(a)$ for all $a \in A$.*

PROOF. In fact, this result holds for all states on A. Indeed, let $\phi_n : A \to M_{k(n)}(\mathbb{C})$ be any sequence of u.c.p. maps such that $\|a\| = \lim \|\phi_n(a)\|$ for all $a \in A$. Consider the following set of states on A:

$$\mathcal{S} = \{\sigma \in S(A) : \exists \sigma_n \in S(M_{k(n)}(\mathbb{C})) \text{ such that } \sigma = \sigma_n \circ \phi_n\}.$$

Since $\|a\| = \lim \|\phi_n(a)\|$ for all $a \in A$ it is easy to see that for every self-adjoint $a \in A$ we have $\|a\| = \sup\{|\sigma(a)| : \sigma \in \mathcal{S}\}$. From the Hahn-Banach theorem it follows that the weak-$*$ closure of $\mathcal{S}$ is the entire state space of A (cf. Lemma 2.8). An argument similar to that given in the proof of Proposition 3.5.1 completes the proof. □

Hence we wondered if one could always replace the *states* in the lemma above with traces by passing to centralizers or some other general argument. Unfortunately this is impossible, in general, though the nuclear case and the case of $C^*(\mathbb{F}_\infty)$ are still open.

COROLLARY 4.3.8. *There exists an exact, residually finite dimensional[13] (hence quasidiagonal) C*-algebra A such that T(A) $\neq$ AT(A) (hence T(A) $\neq$ AT(A)$_{\text{QD}}$).*

PROOF. Since $C^*_r(\mathbb{F}_2)$ is exact and every exact C*-algebra is a quotient of an exact, residually finite dimensional algebra (cf. [**17**]) it follows that we can find an exact, residually finite dimensional C*-algebra A and a trace $\tau \in$ T(A) such that $\pi_\tau(A)''$ is a free group factor. Since free group factors are not hyperfinite, τ is not an amenable trace on A. □

4.4. Type I C*-algebras

Recall from Theorem 4.2.1 that *every trace on a nuclear C*-algebra is a uniform amenable trace*. It may be that this is the strongest approximation property enjoyed by the entire class of nuclear C*-algebras. However, in this section we will show that for type I C*-algebras a much stronger approximation property always holds – every trace is uniform locally finite dimensional.

Our first lemma is rather technical but gives a useful characterization of these traces.

LEMMA 4.4.1. *Let $\tau \in$ T(A) be given. Then $\tau \in$ UAT(A)$_{\text{LFD}}$ if and only if for every finite set $\mathfrak{F} \subset A$ and $\epsilon > 0$ there exists a C*-subalgebra $B \subset A$ and a u.c.p. map $\phi : B \to M_n(\mathbb{C})$ such that $d(\mathfrak{F}, B_\phi) < \epsilon$ and $\|\tau|_B - \text{tr}_n \circ \phi\|_{B^*} < \epsilon$.*

---

[13]A is residually finite dimensional if it has a separating family of finite dimensional representations. Another way of saying this is that A embeds into $\Pi M_{k(n)}(\mathbb{C})$ for some sequence of integers $k(n)$.

PROOF. Evidently we need only show the 'if' part. The obvious approach to proving this lemma would be to simply extend, via Arveson's theorem, the map on B to all of A. However, it does not seem obvious that if $\tilde{\phi} : A \to M_n(\mathbb{C})$ is a u.c.p. extension then we still have $\|\tau - \mathrm{tr}_n \circ \tilde{\phi}\|_{A^*} < \epsilon$ and hence our proof will take another route.

So, assume $\mathfrak{F} \subset A$ and $\epsilon > 0$ are given and choose $B \subset A$ and $\phi : B \to M_n(\mathbb{C})$ such that $d(\mathfrak{F}, B_\phi) < \epsilon$ and $\|\tau|_B - \mathrm{tr}_n \circ \phi\|_{B^*} < \epsilon$.

Let $P_\tau \in A^{**}$ (resp. $P_{\mathrm{tr}_n \circ \phi} \in B_\phi^{**} \subset B^{**} \subset A^{**}$) be the central cover (cf. [**61**]) of the GNS representation of τ (resp. the GNS representation of $\mathrm{tr}_n \circ \phi|_{B_\phi}$). Let $Q = P_\tau P_{\mathrm{tr}_n \circ \phi} \in A^{**}$. Then Q commutes with B_ϕ ($\subset A^{**}$) and QB_ϕ is a finite dimensional subalgebra of $QA^{**}Q \subset P_\tau A^{**} \cong \pi_\tau(A)''$ (since $P_{\mathrm{tr}_n \circ \phi}B_\phi$ is finite dimensional). Letting τ^{**} denote the normal extension of τ to A^{**}, note that $\tau^{**}(Q) = \tau^{**}(P_{\mathrm{tr}_n \circ \phi}) = \tau^{**}|_{B_\phi^{**}}(P_{\mathrm{tr}_n \circ \phi}) = (\tau|_{B_\phi})^{**}(P_{\mathrm{tr}_n \circ \phi}) > 1 - \epsilon$ since $\|\tau|_{B_\phi} - \mathrm{tr}_n \circ \phi|_{B_\phi}\|_{B_\phi^*} < \epsilon$. Let $E : QA^{**}Q \to QB_\phi$ be a conditional expectation which preserves the tracial state θ on $QA^{**}Q$ defined by $QxQ \mapsto \frac{1}{\tau^{**}(Q)}\tau^{**}(QxQ)$. Then define $\Phi : A \to QB_\phi$ by $\Phi(a) = E(QaQ)$. Note that $B_\phi \subset A_\Phi$ and hence the proof is complete once one observes that

$$|\tau(a) - \theta \circ \Phi(a)| = |\tau^{**}(Q^\perp a) + \tau^{**}(Qa) - \frac{1}{\tau^{**}(Q)}\tau^{**}(Qa)| \leq (\epsilon^{1/2} + \frac{\epsilon}{1-\epsilon})\|a\|,$$

and then applies Lemma 2.5 to get from QB_ϕ to a full matrix algebra. □

The following lemma of Huaxin Lin provides the key first step to proving that all traces on type I C*-algebras are uniform locally finite dimensional.

LEMMA 4.4.2. *(cf. [**53**, Lemma 4.7]) Let B be a subhomogeneous C^*-algebra (i.e. assume the dimension of every irreducible representation of B is uniformly bounded). Then, for every trace $\tau \in \mathrm{T}(B)$, finite set $\mathfrak{F} \subset B$ and $\epsilon > 0$ there exists an ideal $I \subset B$, a trace $\gamma \in \mathrm{T}(B/I)$ and a unital, finite dimensional subalgebra $C \subset B/I$ such that (1) $\|\tau - \gamma \circ \sigma\|_{B^*} < \epsilon$ and (2) $d(\sigma(\mathfrak{F}), C) < \epsilon$, where $\sigma : B \to B/I$ is the canonical quotient mapping.*

LEMMA 4.4.3. *Let $\tau \in \mathrm{T}(A)$ be a trace and assume that the weak closure of the associated GNS representation of A is a type I von Neumann algebra. Then $\tau \in \mathrm{UAT}(A)_{\mathrm{LFD}}$.*

PROOF. Fix a finite set $\mathfrak{F} \subset A$ and $\epsilon > 0$. Let $\pi_\tau : A \to B(L^2(A, \tau))$ be the GNS representation arising from τ. Since $\pi_\tau(A)''$ is a type I von Neumann algebra we have two cases to consider.

Case 1: Assume $\pi_\tau(A)''$ is isomorphic to

$$\prod_{i=1}^{k} M_{n(i)}(\mathbb{C}) \otimes L^\infty(X_i, \mu_i)$$

for some probability spaces (X_i, μ_i) and some natural number k. Let $B = \pi_\tau(A)$ and $\tilde{\mathfrak{F}} = \pi_\tau(\mathfrak{F}) \subset B$. Then B is subhomogeneous and by Lin's lemma we can find an ideal $I \subset B$, trace $\gamma \in \mathrm{T}(B/I)$ and unital, finite dimensional subalgebra $C \subset B/I$ such that $\|\tau - \gamma \circ \sigma\|_{B^*} < \epsilon$ and $d(\sigma(\tilde{\mathfrak{F}}), C) < \epsilon$. Note that if $D = (\sigma \circ \pi_\tau)^{-1}(C) \subset A$ is the pullback of C then $d(\mathfrak{F}, D) < \epsilon$. Note also that $\|\tau|_D - \gamma \circ \sigma \circ \pi_\tau|_D\|_{D^*} < \epsilon$ in this situation. Applying Lemma 4.4.1 to the subalgebra $D \subset A$ and *-homomorphism $\sigma \circ \pi_\tau : D \to C$ we get the result in this case. (Strictly speaking one should apply

Lemma 2.5 to get from C to a full matrix algebra, but we have seen this type of argument numerous times already.)

Case 2: The only other possibility is that the weak closure of $\pi_\tau(A)$ is isomorphic to
$$\prod_{i=1}^\infty M_{n(i)}(\mathbb{C}) \otimes L^\infty(X_i, \mu_i).$$
However, in this case
$$\tau = \sum_{i=1}^\infty \theta_i \tau_i$$
where θ_i are positive real numbers which sum up to one and τ_i is gotten by restricting τ to the ith summand of $\prod_{i=1}^\infty M_{n(i)}(\mathbb{C}) \otimes L^\infty(X_i, \mu_i)$ (and renormalizing to get a state).[14] However, each of the tracial functionals
$$\eta_k = \sum_{i=1}^k \theta_i \tau_i$$
falls under Case 1 and the η_k's converge in norm to τ. In other words, τ is a norm limit of elements of $\mathrm{UAT}(A)_{\mathrm{LFD}}$ and since this set is norm closed it follows that $\tau \in \mathrm{UAT}(A)_{\mathrm{LFD}}$ as well. □

COROLLARY 4.4.4. *If A is an inductive limit of type I C^*-algebras then $\mathrm{T}(A) = \mathrm{UAT}(A)_{\mathrm{LFD}}$. In particular, every trace on a type I C^*-algebra is uniform locally finite dimensional.*

PROOF. Let $\tau \in \mathrm{T}(A)$ be given and assume A is the norm closure of type I subalgebras $\{A_i\}$. By the previous lemma we have that $\tau|_{A_i}$ is uniform locally finite dimensional and hence from Lemma 4.4.1 it follows that $\tau \in \mathrm{UAT}(A)_{\mathrm{LFD}}$. □

Note, of course, that one really does not need an inductive limit decomposition in the previous corollary. It suffices to be "locally type I" in the sense that every finite set is almost contained in a type I subalgebra (though the type I subalgebras need not be increasing).

4.5. Tracially AF C*-algebras

This section is devoted to Huaxin Lin's class of tracially AF algebras [51]. This class of C*-algebras is similar to the class of Popa algebras, however Lin's work on classifying such algebras has been a breakthrough in the classification program and proving that particular C*-algebras are tracially AF now immediately yields new classification results.

Our first goal is to show that tracial approximation properties give a very simple characterization of these algebras in many instances. This was also recognized by Lin in [53] where he introduced a notion of 'approximately AC trace' (cf. [53, Definition 3.1]) and used this to characterize certain tracially AF algebras. The present approach, however, has the advantage of being conceptually and technically easier to digest. Moreover, our definitions do not presuppose an inductive limit type structure as is required in [53, Definition 3.1].

The following theorem summarizes several of Lin's results and explains the 'tracially AF' terminology (cf. [52, Theorems 6.9, 6.11 and 6.13]).

---

[14]This is because of normality of the induced vector trace in the GNS representation.

THEOREM 4.5.1. *Let A be a simple C^*-algebra. Then A is tracially AF if and only if A has real rank zero, stable rank one, weakly unperforated K-theory[15] and for every finite subset $\mathfrak{F} \subset A$ and $\epsilon > 0$ there exists a finite dimensional subalgebra $B \subset A$ with unit e such that:*

(1) $\|[x,e]\| < \epsilon$, for all $x \in \mathfrak{F}$.
(2) $d(e\mathfrak{F}e, B) < \epsilon$.
(3) $\tau(e) > 1 - \epsilon$ for all $\tau \in \mathrm{T}(A)$.

As mentioned above, we are primarily interested in knowing which C*-algebras are tracially AF. We remind the reader that Popa's work shows that every simple, quasidiagonal C*-algebra with real rank zero is 'almost' tracially AF (more precisely, satisfies the approximation property with (1) and (2), but not necessarily (3), above). The point of this section is that if one happens to know that the traces on A are sufficiently well behaved then it can be shown that A is tracially AF (i.e. we can also arrange condition (3)). The difficulty is that condition (3) is global (i.e. must hold for all traces) and hence if one only knows approximation properties of particular traces then one is forced to add restrictions on the size of $\mathrm{T}(A)$ so that a constructive procedure can be carried out. The following lemma, which for all intents and purposes is due to Lin, makes this more precise.

LEMMA 4.5.2. *Assume that $\mathrm{T}(A)$ is $\|\cdot\|_{A^*}$-separable and every hereditary subalgebra $A_0 \subset A$ has the following (tracially local) approximation property: For every trace $\tau \in \mathrm{T}(A)$, finite subset $\mathfrak{F} \subset A_0$ and $\epsilon > 0$ there exists a finite dimensional subalgebra $B \subset A_0$ with unit e such that:*

(1) $\|[x,e]\| < \epsilon$, for all $x \in \mathfrak{F}$.
(2) $d(e\mathfrak{F}e, B) < \epsilon$.
(3) $\tau(e) > (1-\epsilon)\|\tau|_{A_0}\|$.

Then it follows that for every finite subset $\mathfrak{F} \subset A$ and $\epsilon > 0$ there exists a finite dimensional subalgebra $B \subset A$ with unit e such that:

(1') $\|[x,e]\| < \epsilon$, for all $x \in \mathfrak{F}$.
(2') $d(e\mathfrak{F}e, B) < \epsilon$.
(3') $\tau(e) > 1 - \epsilon$ for all $\tau \in \mathrm{T}(A)$.

In other words, if one assumes that $\mathrm{T}(A)$ is $\|\cdot\|_{A^*}$-separable and happens to know that for each trace one can find a "large" subalgebra with the right tracially AF properties then, in fact, it is possible to find a subalgebra which is large in all traces simultaneously.[16]

PROOF. This lemma is basically an abstraction of the proof of [53, Theorem 4.13]. Hence we will only sketch the main idea.

Fix a finite subset $\mathfrak{F} \subset A$ and $\epsilon > 0$. Let $\{\tau_i\} \subset \mathrm{T}(A)$ be a *norm* dense sequence. Applying the assumed approximation property to $(\mathfrak{F}, \tau_1, \epsilon/2)$ we can find B_1 with unit e_1 such that

---

[15]Recall that real rank zero means the self-adjoint elements with finite spectrum are dense in the set of all self-adjoints, stable rank one means the invertible elements are dense in the whole algebra and weakly unperforated K-theory means that if $x \in K_0(A)$ and $\tau(x) > 0$ for all $\tau \in \mathrm{T}(A)$ then $x > 0$.

[16]It would be very nice if one could remove the assumption of $\|\cdot\|_{A^*}$-separability from this result – not just for aesthetic reasons but also because it would immediately imply significant new classification results. While we believe this should be possible it has resisted our best efforts and those of a few other colleagues we have spoken to.

(a) $\|[x, e_1]\| < \epsilon/2$, for all $x \in \mathfrak{F}$,
(b) $d(e_1 \mathfrak{F} e_1, B_1) < \epsilon/2$,
(c) $\tau_1(e_1) > 1 - \epsilon/2$.

Now we look at the hereditary subalgebra which is orthogonal to B_1 (i.e. $e_1^\perp A e_1^\perp$) and apply the assumed approximation property to $(e_1^\perp \mathfrak{F} e_1^\perp, \tau_2, \epsilon/8)$. We get a finite dimensional subalgebra $B_2 \subset e_1^\perp A e_1^\perp$ with unit e_2 satisfying the stated conditions. The key observation is that the finite dimensional subalgebra $B_1 \oplus B_2$ has the following property:

(d) $\|[x, e_1 \oplus e_2]\| < \epsilon/2 + \epsilon/8$, for all $x \in \mathfrak{F}$,
(e) $d([e_1 \oplus e_2]\mathfrak{F}[e_1 \oplus e_2], B_1 \oplus B_2) < \epsilon/2 + \epsilon/4$,
(f) $\tau_2(e_1 \oplus e_2) > 1 - \epsilon/8$.

By induction we can produce a sequence of finite dimensional subalgebras C_n[17] with units f_n such that $f_1 \leq f_2 \leq f_3 \leq \cdots$ and

(g) $\|[x, f_n]\| < \epsilon$, for all $x \in \mathfrak{F}$,
(h) $d(f_n \mathfrak{F} f_n, C_n) < \epsilon$,
(i) $\tau_n(f_n) > 1 - \frac{\epsilon}{2^n}$.

The only thing left to verify is that $\tau(f_n) \to 1$ for every $\tau \in \mathrm{T}(A)$ because if this is the case then Dini's theorem applies (identifying $\{f_n\}$ with an *increasing* sequence of functions on the compact metric space $\mathrm{T}(A)$) and we conclude that the convergence is uniform; i.e. there exists some large n such that $\tau(f_n) > 1 - \epsilon$ for all $\tau \in \mathrm{T}(A)$. This is where the assumption of $\|\cdot\|_{A^*}$-separability of $\mathrm{T}(A)$ comes in. Indeed, we may assume that the sequence $\{\tau_i\} \subset \mathrm{T}(A)$ has infinite multiplicity (i.e. every trace in the sequence appears infinitely many times in the sequence) and thus condition (i) ensures that $\tau(f_n) \to 1$ on a *norm* dense subset of $\mathrm{T}(A)$. However this implies pointwise convergence to one on all of $\mathrm{T}(A)$ and we are done. $\square$

LEMMA 4.5.3. *Let $\tau \in \mathrm{UAT}(A)_{\mathrm{LFD}}$ and a projection $p \in A$ be given. Then the tracial state $\frac{1}{\tau(p)}\tau|_{pAp}$ on pAp is also uniform locally finite dimensional.*

PROOF. Let $p \in \mathfrak{F} \subset pAp$ and $\epsilon > 0$ be given. Let $\delta > 0$ be much smaller (to be determined later) than ϵ and let $\phi : A \to M_k(\mathbb{C})$ be a u.c.p. map such that $\|\mathrm{tr}_k \circ \phi - \tau\| < \tau(p)\delta$ and $\mathfrak{F}$ is δ-contained in the multiplicative domain A_ϕ of ϕ. Since p is a projection it follows from some standard K-theoretic results that we can find a projection $q \in A_\phi$, which is very close to p, and a unitary $u \in A$, which is very close to the identity, such that $upu^* = q$. Hence defining $\psi : pAp \to \phi(q)M_k(\mathbb{C})\phi(q) \cong M_l(\mathbb{C})$ (for some $l \leq k$) by the formula $\psi(pap) = \phi(upapu^*)$ we see that $\mathfrak{F}$ is nearly contained in the multiplicative domain of ψ (since $q \in A_\phi$ and u is close in norm to the identity, $u^* q A_\phi q u$ is in the multiplicative domain of ψ and almost contains $\mathfrak{F}$) and $\mathrm{tr}_l \circ \psi$ is very close in norm (within δ actually) to $\frac{1}{\tau(p)}\tau|_{pAp}$. $\square$

We will need one more lemma before coming to our characterization of tracially AF algebras in terms of tracial approximation properties. The following result is taken from the work of Huaxin Lin (cf. [**51**, proof of Theorem 5.3]) but we include those ideas in the proof which are no so easy to extract from [**51**].

---

[17]$C_1 = B_1$, $C_2 = B_1 \oplus B_2, \ldots$.

LEMMA 4.5.4. *Assume that A has real rank zero, $1_A \in D \subset A$ is a C^*-subalgebra which contains an ideal $I \triangleleft D$ such that I is a hereditary subalgebra of A and $D/I = C$ is finite dimensional. Then the exact sequence*

$$0 \to I \to D \xrightarrow{\pi} C \to 0$$

has a $*$-homomorphic splitting – i.e. there is a $*$-homomorphism $\sigma : C \to D$ such that $\pi \circ \sigma = id_C$. Moreover, there is an approximate unit of projections $\{q_n\} \subset I$ such that $[q_n, \sigma(C)] = 0$ for all $n \in \mathbb{N}$.

PROOF. Once one knows that projections from C can always be lifted to D then the existence of a $*$-homomorphic splitting follows exactly as in the proof of [**27**, Lemma III.6.2]. However, a nice proof that one can always lift projections can be found in [**51**, Lemma 5.2] and hence our map σ always exists.

That an approximate unit exists which both consists of projections and commutes with $\sigma(C)$ is a bit technical but only relies on fairly standard matrix manipulations. The main ingredients are already contained in the special case that $C \cong M_2(\mathbb{C})$ (where notation becomes much simpler) so we give all the details in this setting and leave the general case to the reader.

So let $B = \sigma(C) \subset A$ and denote by e_1 and e_2 a pair of orthogonal minimal projections in B. Let $u \in B$ be a unitary such $ue_1u^* = e_2$. Consider the ideal $e_1Ie_1 \triangleleft e_1De_1$. Since real rank zero passes to hereditary subalgebras it follows that e_1Ie_1 has real rank zero and hence we can find an approximate unit of projections $\{p_n\} \subset e_1Ie_1$. Evidently $\{up_nu^*\} \subset e_2Ie_2$ is also an approximate unit of projections and hence we can define projections

$$q_n = r_n + p_n + up_nu^* \in I,$$

where $\{r_n\} \subset (e_1+e_2)^\perp I(e_1+e_2)^\perp$ is also an approximate unit of projections. We must verify that these projections form an approximate unit for I but the commutation part is now simple since one easily checks that each q_n commutes with all four of the operators e_1, e_2, ue_1 and e_1u^* (and these four operators are a set of matrix units for B).

In general, if $I \triangleleft D$ is an ideal, $\{s_n\} \subset dId$ is an approximate unit ($d \in D$ is some fixed element) and $x \in I$ is arbitrary then a straightforward calculation using the C^*-identity shows that

$$\|(xd)s_n - xd\| \to 0.$$

Now, our goal is to show that $q_n = r_n + p_n + up_nu^*$ is an approximate unit of I and this essentially reduces to the remark in the preceding sentence. Indeed, we can write each $x \in I$ as a 3×3 matrix with respect to the decomposition $1 = (e_1 + e_2)^\perp + e_1 + e_2$. It then suffices to consider the "off diagonal elements" and show, for example,

$$\|e_1 x(e_1+e_2)^\perp q_n - e_1 x(e_1+e_2)^\perp\| = \|e_1\big(x(e_1+e_2)^\perp r_n - x(e_1+e_2)^\perp\big)\| \to 0.$$

This follows from the general remarks above and hence we leave the remaining details to the reader.

For more general finite dimensional algebras B we first decompose B as a finite direct sum of full matrix algebras and then apply the procedure above underneath each of the summands. It is quite messy to properly write down so we won't attempt it. □

PROPOSITION 4.5.5. *Let A be a simple C^*-algebra with real rank zero, stable rank one, weakly unperforated K-theory and assume that $\mathrm{T}(A)$ is norm separable. Then A is tracially AF if and only if $\mathrm{T}(A) = \mathrm{UAT}(A)_{\mathrm{LFD}}$.*

PROOF. We begin with the 'only if' statement. So assume that A is tracially AF and let $\tau \in \mathrm{T}(A)$, a finite set $\mathfrak{F} \subset A$ and $\epsilon > 0$ be given. Choose a finite dimensional subalgebra $B \subset A$ as in Lin's characterization above and note that there is a conditional expectation $eAe \to B$ which preserves the tracial state $\frac{1}{\tau(e)}\tau|_{eAe}$. Composing this map with the u.c.p. map $A \to eAe$, $a \mapsto eae$ we get a u.c.p. map from A to B and $\mathfrak{F}$ is nearly contained in the multiplicative domain of this map since condition (1) in Theorem 4.5.1 above implies that $a \approx eae \oplus e^\perp a e^\perp$ for all $a \in \mathfrak{F}$ (and clearly $e^\perp A e^\perp$ belongs to the multiplicative domain). Finally, it is easily seen that τ is close (in norm) to $\frac{1}{\tau(e)}\tau(e \cdot e)$ since $\tau(e)$ is close to one.

For the converse, assume that $\mathrm{T}(A) = \mathrm{UAT}(A)_{\mathrm{LFD}}$ and we will show that A has the 'tracially local' approximation property from Lemma 4.5.2. Note that from the proof of that lemma we really only need to consider corners of A (instead of general hereditary subalgebras). Moreover, by the previous lemma, restrictions of traces on A will again be uniform locally finite dimensional and hence we may assume that the hereditary subalgebra is A itself (since the proof will carry over verbatim to corners). Hence we need to show that if $\mathfrak{F} \subset A$ is a finite set, $\tau \in \mathrm{T}(A)$ is arbitrary and $\epsilon > 0$ is given then we can find a finite dimensional subalgebra $B \subset A$ with unit e such that $\tau(e)$ is large, e almost commutes with $\mathfrak{F}$ and cutting down by e almost pushes $\mathfrak{F}$ into B.

By assumption we can find a u.c.p. map $\phi : A \to M_k(\mathbb{C})$ such that $\|\mathrm{tr}_k \circ \phi - \tau\|_{A^*} < \epsilon$ and $\mathfrak{F}$ is nearly contained (within ϵ) in the multiplicative domain A_ϕ of ϕ. Let $J \subset A_\phi$ be the kernel of the $*$-homomorphism $\phi|_{A_\phi}$. We claim that J is a hereditary subalgebra of A and hence inherits real rank zero from A. To see this we assume $0 \leq x \leq y$ and $y \in J$. By positivity we have that $\phi(x) = 0$. The Cauchy-Schwartz inequality for c.p. maps tells us that $0 \leq \phi(\sqrt{x})^2 \leq \phi((\sqrt{x})^2) = 0$ and hence $\phi(\sqrt{x}) = 0$ as well. This shows that $\sqrt{x} \in A_\phi$ and thus x belongs to the multiplicative domain too. It follows that $x \in J$ which shows that J is a hereditary subalgebra of A.

Applying Lemma 4.5.4 to the short exact sequence

$$0 \to J \to A_\phi \to C \to 0,$$

where $C = \phi(A_\phi) \subset M_k(\mathbb{C})$, we can find a finite dimensional C*-subalgebra $B \subset A_\phi$, with unit e, such that ϕ is an isomorphism from B onto C. Moreover, we can find an approximate unit of projections $\{q_n\} \subset J$ which commutes with B and hence is quasicentral in A_ϕ. Borrowing a few more techniques from the proof of [51, Theorem 5.3] will complete the proof. The main new ingredient, however, we wish to explain from the outset. Namely, note that

$$\tau((1 - q_n)e) > 1 - \epsilon$$

for every n since $\|\mathrm{tr}_k \circ \phi - \tau\|_{A^*} < \epsilon$, $(1-q_n)e$ belongs to the unit ball of A and, finally, $\phi((1-q_n)e) = 1$ since $q_n \in J$ and e is a lift of the unit of C.

Hence the proof will be complete as soon as we verify:
(1) $\limsup \|[x, (1-q_n)e]\| < \epsilon$ for all $x \in \mathfrak{F}$.
(2) $\limsup d((1-q_n)ex(1-q_n)e, (1-q_n)eB) < \epsilon$ for all $x \in \mathfrak{F}$.

Indeed, if we are successful in showing this then we can take our desired finite dimensional algebra to be $(1 - q_n)B$ for any sufficiently large n. Since $\mathfrak{F}$ is almost contained in A_ϕ it will be sufficient to replace x above by an element in the multiplicative domain and show that the lim sup's are actually zero in this case. But if $x \in A_\phi$ then

$$\|x(1 - q_n)e - (1 - q_n)ex\| \stackrel{\epsilon}{\approx} \|(1 - q_n)(xe - ex)\| \to \|\phi(xe - ex)\| = 0$$

where we used the fact that $\{q_n\}$ is quasicentral to get the first approximation and the fact that e is a lift of 1 (which is central in the quotient, of course) for the second part. For the assertion $d((1 - q_n)ex(1 - q_n)e, (1 - q_n)eB) \to 0$ we first pick a lift $b \in B$ of $\phi(x) \in C$. A similar argument to the one given above shows that $d((1 - q_n)ex(1 - q_n)e, (1 - q_n)eb) \to 0$ and hence the proof is complete. $\square$

REMARK 4.5.6. This proposition is the analogue of Theorems 3.8 and 4.13 from [53]. Though our proof certainly uses some key ideas from [53] it is also true that the present approach avoids a number of technical difficulties which must be dealt with in Lin's approach. Note that the 'only if' part holds without the separability assumption on the tracial space. Also, the 'if' statement actually generalizes [53, Theorem 4.13] as it can be shown that every 'approximately AC trace' is uniform locally finite dimensional. Indeed, it is not too hard to show that every 'AC trace' is uniform locally finite dimensional and so Lemma 4.4.1 implies that approximately AC traces have this property as well.

Though it won't be needed in what follows we wish to point out that the (annoying) assumption that T(A) is norm separable is basically necessary if one wishes to carry out a constructive procedure as in the proof of Lemma 4.5.2. Moreover, we remind the reader that a constructive procedure is also basically necessary in the category of C*-algebras (i.e. one can't use Zorn's Lemma type maximality arguments – which work so beautifully in von Neumann algebras – since an infinite sum of projections doesn't make sense in a C*-algebra).

PROPOSITION 4.5.7. *Let A be a simple, nuclear C*-algebra with real rank zero, stable rank one and weakly unperforated K-theory. Then the following statements are equivalent:*
  (1) *T(A) is weakly (i.e. $\sigma(A^*, A^{**})$) separable.*
  (2) *T(A) is norm separable.*
  (3) *T(A) has countably many extreme points.*
  (4) *There exists a countable subset $\mathcal{S} \subset T(A)$ with the following property: For every increasing sequence of projections in A, say $p_1 \leq p_2 \leq \cdots$, if one knows that $\tau(p_n) \to 1$ for all $\tau \in \mathcal{S}$ then it follows that $\tau(p_n) \to 1$ for all $\tau \in T(A)$.*

PROOF. The equivalence of (1) and (2) follows from the Hahn-Banach Theorem and the convexity of T(A).

(3) $\implies$ (2). By Choquet theory, for every $\tau \in T(A)$ there exist non-negative real numbers $\{a_i\}$ such that $\sum a_i = 1$ and if we list the extreme points of T(A) as $\{\tau_1, \tau_2, \ldots\}$ then

$$\tau = \sum a_i \tau_i.$$

Evidently this implies T(A) is norm separable.

(2) $\Longrightarrow$ (3). We prove the contrapositive. Note that since extreme points of T(A) yield factors in their GNS representations and, in general, factor representations are either disjoint or equivalent (cf. [**61**, 3.8.13]) it follows (using the fact that finite factors have a unique trace) that for two extreme traces τ_1, τ_2 we either have that $\tau_1 = \tau_2$ or $\|\tau_1 - \tau_2\|_{A^*} = 2$. Hence if T(A) has uncountably many extreme points we see that it can't be norm separable.

(2) $\Longrightarrow$ (4). Any norm dense subset has the property described in condition (4).

(4) $\Longrightarrow$ (2). We again prove the contrapositive. So assume T(A) is not separable in norm and let $\pi = \oplus_{\tau \in T(A)} \pi_\tau : A \to B(\oplus_{\tau \in T(A)} L^2(A, \tau))$ be the direct sum of all GNS representations arising from traces on A. Let $M = \pi(A)''$ and note that M is a finite von Neumann algebra and every trace on A naturally extends to a weakly continuous trace on M. Hence the predual of M is not separable (since we assume T(A) is not separable). Thus M has no faithful representation on a separable Hilbert space.

Now let $\mathcal{S} \subset T(A)$ be any countable subset and we will show that there exists an increasing sequence of projections $p_1 \leq p_2 \leq \cdots$ such that $\tau(p_n) \to 1$ for all $\tau \in \mathcal{S}$ but such that there exists some $\gamma \in T(A)$ for which $\gamma(p_n)$ does not tend to one. The first step in constructing such projections is to construct an increasing sequence of positive elements from the unit ball of A with this property. This will follow from Pedersen's Up-Down Theorem (cf. [**61**, Theorem 2.4.3]).

Let $\sigma_{\mathcal{S}} : M \to B(\oplus_{\tau \in \mathcal{S}} L^2(A, \tau))$ be the natural restriction of M to the sum of the GNS spaces coming from traces in $\mathcal{S}$. Note that $\oplus_{\tau \in \mathcal{S}} L^2(A, \tau)$ is a separable Hilbert space and hence σ can't be faithful. Thus there exists a central projection $q \in M$ such that qM is naturally isomorphic to $\sigma(M)$. Evidently we can then find a trace $\gamma \in T(A)$ such that $\gamma(q) = 0$ (γ also denoting the natural extension to M). Note also that $\tau(q) = 1$ for all $\tau \in \mathcal{S}$. Now let $\sigma_{\tilde{\mathcal{S}}} : M \to B(\oplus_{\tau \in \tilde{\mathcal{S}}} L^2(A, \tau))$. We will identify qM with a (nontrivial) direct summand of $\sigma_{\tilde{\mathcal{S}}}(M)$. Note that since A is simple, $\sigma_{\tilde{\mathcal{S}}}(A) \cong A$. At this point what we have arranged is: There exists a representation of A, $\sigma : A \to B(H)$, on a separable Hilbert space, such that every trace in $\tilde{\mathcal{S}}$ naturally extends to a weakly continuous trace on $\sigma(A)''$ and there exists a projection $q \in \sigma(A)''$ such that $\gamma(q) = 0$ while $\tau(q) = 1$ for all $\tau \in \mathcal{S}$ (these are the abstract properties we really need).

Since $A \cong \sigma(A)$ is weakly dense in $\sigma(A)''$ we can apply Pedersen's Up-Down Theorem to find a sequence of positive elements from the unit ball of A, say $\{a_i\}$, such that (a) $0 < a_1 < a_2 < \cdots$ and (b) $a_i \to Q \in \sigma(A)''$ (weakly) (c) $Q \geq q$ and (d) $\gamma(Q) < 1$ (actually, as close to zero as you like). From (c) and the fact that $\|Q\| = 1$ it follows that $\tau(Q) = 1$ for all $\tau \in \mathcal{S}$. Hence, from (b) and (d), we see that $\tau(a_n) \to 1$ for all $\tau \in \mathcal{S}$ while $\lim \gamma(a_n) < 1$.

Our final task is to replace the sequence of positive elements $\{a_i\}$ above with an increasing sequence of projections which have the same property. To do this we need to recall the following K-theoretic facts about simple, nuclear C*-algebras with real rank zero, stable rank one and weakly unperforated K-theory (cf. [**11**, Corollary 6.9.2 and Theorem 6.9.3]; we also need Haagerup's result that quasitraces on nuclear algebras are actually traces [**36**]): (1) For two projections p and q in (matrices over) A, p is equivalent to a proper subprojection of q if and only if $\tau(p) < \tau(q)$ for all $\tau \in T(A)$; (2) If $f : T(A) \to [0, 1]$ is any continuous, affine function and $\epsilon > 0$ is given then there exists a projection $p \in A$ such that $|f(\tau) - \tau(p)| < \epsilon$ for all $\tau \in T(A)$

(this is essentially a special case of the fact that $K_0(A)$ is dense in $Aff(\mathrm{T}(A))$ together with the strict ordering fact from (1)). With these results in hand it is a simple matter to produce a sequence of projections $\{p_n\}$ in A such that $\tau(p_n) \to 1$ for all $\tau \in \mathcal{S}$ while $\lim \gamma(p_n) < 1$ and, moreover, such that $0 < [p_1] < [p_2] < \cdots$ (in $K_0(A)$) since each of the positive elements a_i naturally defines a positive affine function on $\mathrm{T}(A)$ and there is a uniform gap between each of these functions (where we can stick a projection by (2)) since A is simple (hence every trace is faithful). The final step is to recall that in a C*-algebra with stable rank one, two projections are (Murray-von Neumann) equivalent if and only if they are unitarily equivalent. Hence we can find unitaries $u_n \in A$ such that $p_1 < u_2 p_2 u_2^* < u_3 p_3 u_3^* < \cdots$. □

REMARK 4.5.8. Note that the equivalence of (1), (2) and (3) above holds for arbitrary C*-algebras.

COROLLARY 4.5.9. *Let A be a (separable) simple, nuclear C^*-algebra with real rank zero, stable rank one, weakly unperforated K-theory and such that $\mathrm{T}(A)$ is not norm separable. Then for every countable, weak-∗ dense subset $\mathcal{S} \subset \mathrm{T}(A)$ there exists a sequence of projections $p_1 \leq p_2 \leq \cdots$ such that $\tau(p_n) \to 1$ for all $\tau \in \mathcal{S}$ but $\lim \gamma(p_n) < 1$ for some $\gamma \in \mathrm{T}(A)$.*

CHAPTER 5

Finite representations

Now that we have spent some time discussing approximation properties of traces we will turn to another aspect of finite representation theory. A typical goal in the representation theory of groups is to classify all *irreducible* representations of a given group – i.e. factor representations of type I. Our goal in these notes is quite different, however, as we wish to study what sort of II_1-factors can arise as GNS representations of a given C*-algebra or class of C*-algebras. We believe it is premature to ask for classification results in the realm of type II representations[1] and hence will stick to the more modest goal of constructing as many different type II_1 representations as possible.

5.1. II_1-factor representations of some universal C*-algebras

This section will be concerned with the II_1-factor representation theory of a few examples of universal C*-algebras. More precisely, we will consider one of the infinite (universal) free product C*-algebras

$$\operatornamewithlimits{\text{\Large$*$}}_1^\infty C(\mathbb{T}), \operatornamewithlimits{\text{\Large$*$}}_1^\infty C([-1,1]) \text{ or } \operatornamewithlimits{\text{\Large$*$}}_1^\infty C([0,1])^2$$

and study their II_1-factor representations. The arguments involved are the same in all three cases so *throughout this section we will let A denote any one of the C*-algebras above.*[3]

Since every C*-algebra arises as a quotient of A it is not hard to see that for every von Neumann algebra $M \subset B(H)$ there is a representation $\pi : A \to M \subset B(H)$ such that $\pi(A)'' = M$. A less obvious fact is that in many cases π can be taken to be *injective*.

PROPOSITION 5.1.1. *Let M be a II_1-factor. There exists a faithful trace τ on A such that $\pi_\tau(A)'' \cong M$.*

PROOF. It suffices to construct a *-monomorphism $A \hookrightarrow M$ with weakly dense range. Indeed, uniqueness of GNS representations implies that the (unique, faithful) trace on M, restricted to $A \subset M$, is the τ we are after.

The construction, though a bit technical, really amounts to some universal trickery and is not particularly deep. We first need to write A as an inductive limit

---

[1] Indeed, this is probably hopeless except in trivial cases.

[2] In other words, the universal C*-algebra generated by a countable number of unitaries, contractive self-adjoints or, respectively, contractive positive operators.

[3] In fact, all the results presented here are valid for any C*-algebra A with the following properties: (1) A is isomorphic to $\operatornamewithlimits{\text{\Large$*$}}_1^\infty A$, (2) A is residually finite dimensional and (3) every (separable, unital) C*-algebra arises as a quotient of A.

of free products of itself. That is, we define
$$A_1 = A, A_2 = A_1 * A, \ldots, A_n = A_{n-1} * A, \ldots,$$
where $*$ denotes the full (i.e. universal) free product (with amalgamation over the scalars).

Letting B denote the inductive limit of the sequence $A_1 \to A_2 \to \cdots$ it is easy to see (by universal considerations) that $A \cong B$.[4] Since A is residually finite dimensional (see [**27**] for the case of $*_1^\infty C(\mathbb{T}) \cong C^*(\mathbb{F}_\infty)$ – the other two cases have a similar, but easier, proof) we can find a sequence of integers $\{k(n)\}$ and a unital $*$-monomorphism $\sigma : B \hookrightarrow \Pi M_{k(n)}(\mathbb{C})$. Note that we may naturally identify each A_i with a subalgebra of B and hence, restricting σ to this copy of A_i, get an injection of A_i into $\Pi M_{k(n)}(\mathbb{C})$.

To construct the desired embedding of B into M, it suffices to prove the existence of a sequence of unital $*$-homomorphisms $\rho_i : A_i \to M$ with the following properties:

(1) Each ρ_i is injective.
(2) $\rho_{i+1}|_{A_i} = \rho_i$, where we identify A_i with the 'left side' of $A_i * A = A_{i+1}$.
(3) The (increasing) union of $\{\rho_i(A_i)\}$ is weakly dense in M.

To this end, we first choose an increasing sequence of projections $p_1 \leq p_2 \leq \cdots$ from M such that $\tau_M(p_i) \to 1$. Then define orthogonal projections $q_2 = p_2 - p_1, q_3 = p_3 - p_2, \ldots$ and consider the II$_1$-factors $Q_j = q_j M q_j$ for $j = 2, 3, \ldots$. As is well known and not hard to construct, we can, for each $j \geq 2$, find a unital embedding $\Pi M_{k(n)}(\mathbb{C}) \hookrightarrow Q_j \subset M$ and thus we get a sequence of (orthogonal) embeddings $B \hookrightarrow \Pi M_{k(n)}(\mathbb{C}) \hookrightarrow Q_j \subset M$ which will be denoted by σ_j.

We are almost ready to construct the ρ_i's. Indeed, for each $i \in \mathbb{N}$ let $\pi_i : A \to p_i M p_i$ be a (not necessarily injective!) $*$-homomorphism with weakly dense range. We then define ρ_1 as

$$\rho_1 = \pi_1 \oplus \left(\bigoplus_{j \geq 2} \sigma_j|_{A_1} \right) : A_1 \hookrightarrow p_1 M p_1 \oplus \left(\Pi_{j \geq 2} Q_j \right) \subset M.$$

Note that this is a unital $*$-monomorphism from A_1 into M (since each σ_j is already faithful on all of B). Now define a $*$-homomorphism $\theta_2 : A_2 = A_1 * A \to p_2 M p_2$ as the free product of the $*$-homomorphisms $A_1 \to p_2 M p_2$, $x \mapsto p_2 \rho_1(x) p_2$, and $\pi_2 : A \to p_2 M p_2$. We then put

$$\rho_2 = \theta_2 \oplus \left(\bigoplus_{j \geq 3} \sigma_j|_{A_2} \right) : A_2 \hookrightarrow p_2 M p_2 \oplus \left(\Pi_{j \geq 3} Q_j \right) \subset M.$$

Note that $\rho_2|_{A_1} = \rho_1$. Hopefully it is now clear how to proceed. In general, we construct a map (whose range is dense in $p_{n+1} M p_{n+1}$) $\theta_{n+1} : A_{n+1} = A_n * A \to p_{n+1} M p_{n+1}$ as the free product of the cutdown (by p_{n+1}) of ρ_n and π_{n+1}. This map need not be faithful and hence we take a direct sum with $\oplus_{j \geq n+2} \sigma_j|_{A_{n+1}}$ to remedy this deficiency. It is then easy to see that these maps have all the required properties and hence the proof is complete. $\square$

REMARK 5.1.2. While it is not true that every infinite dimensional von Neumann algebra contains a weakly dense copy of A (e.g. abelian or, more generally,

---

[4] B is also the universal C*-algebra generated by countably many operators of the same type as those that generate A.

homogeneous von Neumann algebras of finite type can't contain A) it is true that every infinite dimensional factor contains a weakly dense copy of A. A careful inspection of the proof shows that one really only needs smaller and smaller corners each of which contains a copy of $\Pi M_{k(n)}(\mathbb{C})$ for the proof above to go through.

5.2. Elliott's intertwining argument for II$_1$-factors

This section contains a fairly straightforward adaptation of Elliott's celebrated approximate intertwining argument which has been so successful in the classification program. It has nothing to do with finite representation theory per se but will be needed in the next section when we study factor representations of Popa algebras. Though usually done in the setting of C*-algebras we will need Elliott's argument in the setting of von Neumann algebras. While not the most general possible form, the following version is more than sufficient for our purposes. The set-up is as follows.

Assume that $M \subset B(L^2(M, \tau))$ and $N \subset B(L^2(N, \gamma))$ are von Neumann algebras acting standardly, with faithful, normal tracial states τ and, respectively, γ. Let $X_1 \subset X_2 \subset \ldots \subset M$ and $Y_1 \subset Y_2 \subset \ldots \subset N$ be (not necessarily unital) C*-subalgebras such that $\cup X_i$ is weakly dense in M and $\cup Y_i$ is weakly dense in N. Further assume that we have c.p. maps $\alpha_n : Y_n \to X_n$, $\beta_n : X_n \to Y_{n+1}$, which are contractive both with respect to the operator norms and the 2-norms coming from τ and γ, and finite subsets $\Lambda_i \subset X_i$ and $\Omega_i \subset Y_i$ with the following properties:

(1) $\Lambda_i \subset \Lambda_{i+1}$, $\Omega_i \subset \Omega_{i+1}$, for all $i \in \mathbb{N}$, and the linear spans of $\cup \Lambda_i$ and $\cup \Omega_i$ are norm dense in $\cup X_i$ and $\cup Y_i$, respectively, and hence weakly dense in M and N, respectively. To simplify things, we will also assume that $x_1, x_2 \in \Lambda_i \implies x_1 x_2 \in \Lambda_{i+1}$ and, similarly, that Ω_{i+1} contains the product of any pair of elements from Ω_i.
(2) $\alpha_i(\Omega_i) \subset \Lambda_i$ and $\beta_i(\Lambda_i) \subset \Omega_{i+1}$ for all $i \in \mathbb{N}$.
(3) Both $\{\alpha_i\}$ and $\{\beta_i\}$ are weakly asymptotically multiplicative. That is, $\|\alpha_i(y_1 y_2) - \alpha_i(y_1)\alpha_i(y_2)\|_{2,\tau} \to 0$, as $i \to \infty$, for all $y_1, y_2 \in \cup Y_i$ and similarly for $\{\beta_i\}$.

THEOREM 5.2.1. *(Elliott's Intertwining) In the setting described above, if it happens that $\|x - \alpha_{n+1} \circ \beta_n(x)\|_{2,\tau} < 1/2^n$ and $\|y - \beta_n \circ \alpha_n(y)\|_{2,\gamma} < 1/2^n$ for all $x \in \Lambda_n$, $y \in \Omega_n$ and all $n \in \mathbb{N}$, then $M \cong N$.*

PROOF. For the proof we will need the following two standard facts: i) every norm bounded sequence which is Cauchy in 2-norm converges (in 2-norm) and ii) on norm bounded subsets, the 2-norm topology is the same as the strong operator topology (since our von Neumann algebras are acting standardly on the L^2 spaces coming from their traces).

Since the α_n's and β_n's are 2-norm contractive, we first claim that for each $y \in \cup \Omega_i$, the sequence $\{\alpha_n(y)\}$ is Cauchy in 2-norm (and similarly for each $x \in \cup \Lambda_i$, $\{\beta_n(x)\}$ is Cauchy in 2-norm). To see this we first fix $m \in \mathbb{N}$ and note that
$$\|\alpha_m(y) - \alpha_{m+1}(y)\|_{2,\tau} \leq \|\alpha_m(y) - \alpha_{m+1} \circ \beta_m(\alpha_m(y))\|_{2,\tau} + \|\alpha_{m+1}(\beta_m \circ \alpha_m(y) - y)\|_{2,\tau}$$
which, in turn, is bounded above by $\frac{1}{2^{m-1}}$. Repeated applications of this inequality shows that for $m \leq n$
$$\|\alpha_m(y) - \alpha_n(y)\|_{2,\tau} \leq \sum_{i=m-1}^{n-2} \frac{1}{2^i}$$

and hence the sequence is Cauchy as claimed. Evidently the sequences $\{\alpha_n(y)\}$ and $\{\beta_n(x)\}$ are still Cauchy when y and x are taken from the linear spans of $\cup \Omega_i$ and $\cup \Lambda_i$.

Since the α_n's and β_n's are norm contractive, it follows that for each y in the linear span of $\cup \Omega_i$, the sequence $\{\alpha_n(y)\}$ is convergent in M (and similarly for all $x \in span(\cup \Lambda_i)$). Hence we can define linear maps $\Phi : span(\cup \Lambda_i) \to N$ and $\Psi : span(\cup \Omega_i) \to M$ by $\Phi(x) = \lim \beta_n(x)$ and $\Psi(y) = \lim \alpha_n(y)$. Note that Φ and Ψ are contractive with respect to both the operator norms and the 2-norms. This implies that Φ and Ψ can be (uniquely) extended to the norm closures of $span(\cup \Lambda_i)$ and $span(\cup \Omega_i)$ (which are weakly dense C*-algebras, by condition (1) above) and, moreover, that these extensions are 2-norm contractive as well. Now one uses Kaplansky's density theorem (and the fact that our extensions are still norm and 2-norm contractive) to extend beyond these weakly dense C*-subalgebras to all of M and N. (i.e. For each $x \in M$ we take a norm bounded sequence $\{x_n\}$ from the norm closure of $span(\cup \Lambda_i)$ which converges to x in 2-norm. The image of $\{x_n\}$ in N is then a norm bounded sequence which is Cauchy in 2-norm and hence we map x to the (2-norm) limit of this sequence.)

To save notation, we will also let $\Phi : M \to N$ and $\Psi : N \to M$ denote the maps constructed in the previous paragraph. Note that these maps are 2-norm contractive and linear. They are also *-preserving for if $x_n \to x$ (in 2-norm) then it follows that $x_n^* \to x^*$ as well (i.e. the strong and strong-* topologies agree on bounded subsets of a tracial von Neumann algebra since $\|x\|_{2,\tau} = \|x^*\|_{2,\tau}$). Since the maps α_n and β_n preserve adjoints by assumption it follows that the maps Φ and Ψ preserve adjoints as well.

It is easy to check that Φ and Ψ are mutual inverses on the spans of $\cup \Lambda_i$ and $\cup \Omega_i$. By 2-norm contractivity it follows that they are mutual inverses on all of M and N. Hence we only have to observe that both Φ and Ψ are multiplicative on M and N. Since multiplication is continuous, on bounded sets, in the 2-norm, it is not hard to check that Φ and, respectively, Ψ are multiplicative on the norm closures of the linear spans of $\cup \Lambda_i$ and, respectively, $\cup \Omega_i$. Finally, another application of Kaplansky's density theorem and a standard interpolation argument allow one to deduce multiplicativity on all of M and N. $\square$

5.3. II_1-factor representations of Popa Algebras

The main motivation for Popa's work in [67] was to try to understand the relationship between quasidiagonality and nuclearity. Indeed, in [67, pg. 157] Popa asked whether every Popa algebra with unique trace is necessarily nuclear. More generally, he asked in [67, Remark 3.4.2] whether the hyperfinite II_1-factor R was the *only* II_1-factor which could arise from a GNS representation of a Popa algebra. However counterexamples to Popa's first question were constructed by Dadarlat in [26]. Interestingly enough, though, in [26] Dadarlat constructs nonnuclear tracially AF algebras and hence all of their II_1-factor representations are hyperfinite (even though the algebras are not nuclear). Support in favor of Popa's second question was provided in [67, Remark 3.4.2] where Popa proved that if a factorial trace is in $UAT(A)_{QD}$ then it gives the hyperfinite II_1-factor (compare with Theorem 3.2.2 – see also the Introduction).

In this section we will show that Popa's second question also has a negative answer. In fact, we will show that there is a universal Popa algebra for the class of

McDuff II$_1$-factors (meaning there exists a Popa algebra A with the property that every II$_1$-factor of the form $R\bar{\otimes}M$ appears as the GNS representation of A with respect to some trace – see Theorem 5.3.3). On the other hand, we also observe that our results on tracial approximation properties yield an affirmative answer to Popa's question in one non-trivial case; if A is a locally reflexive Popa algebra with unique trace τ then $\pi_\tau(A)'' \cong R$ (see Theorem 5.3.2).

We now turn to the main technical result which will allow us to construct a wide variety of II$_1$-factor representations of Popa algebras. The informed reader will note that every aspect of this result can be traced back to the classification program. Indeed, we will adapt the inductive limit techniques of Dadarlat [**26**] to construct new Popa algebras and use Elliott's intertwining argument to understand their GNS representations.

In the following theorem $\mathfrak{C}$ will denote some collection of C*-algebras which is closed under increasing unions (i.e. inductive limits with injective connecting maps) and tensoring with finite dimensional matrix algebras. A C*-algebra E is called *residually finite dimensional* if E has a separating family of finite dimensional representations (i.e. for every $0 \neq x \in E$ there exists a *-homomorphism $\pi : A \to M_n(\mathbb{C})$ such that $\pi(x) \neq 0$).

THEOREM 5.3.1. *Let $E \in \mathfrak{C}$ be a residually finite dimensional C^*-algebra. Then there exists a Popa algebra $A \in \mathfrak{C}$ such that for every $\varepsilon > 0$ we can find a *-monomorphism $\rho : E \hookrightarrow A$ with the property that for each trace $\tau \in \mathrm{T}(E)$, there exists a trace $\gamma \in \mathrm{T}(A)$ such that*

(1) *$|\gamma \circ \rho(x) - \tau(x)| < \varepsilon \|x\|$ for all $x \in E$ and,*
(2) *$\pi_\gamma(A)'' \cong R\bar{\otimes}\pi_\tau(E)''.$*

The proof of this result becomes much more transparent once the main idea is understood. Hence we think it is worthwhile to give the idea first and leave the details to the end.

So suppose that E is a residually finite dimensional C*-algebra and $\tau \in \mathrm{T}(E)$. Let $\mathcal{U}$ be some UHF algebra. Then the canonical, unital inclusion $E \hookrightarrow E \otimes \mathcal{U}$, $x \mapsto x \otimes 1$ is honestly trace preserving (in fact, yields an isomorphism of tracial spaces) and the weak closure in any GNS representation is obviously of the form $R\bar{\otimes}\pi_\tau(E)''$. The problem, of course, is that $E \otimes \mathcal{U}$ is not a Popa algebra. So the idea is that we will use an inductive limit construction to get a sequence

$$E \to E \otimes M_{k(1)}(\mathbb{C}) \to E \otimes M_{k(1)}(\mathbb{C}) \otimes M_{k(2)}(\mathbb{C}) \to \cdots,$$

such that the limit is a Popa algebra, but the connecting maps above will be chosen so that when one applies a trace it will (approximately) look like the sequence which yields $E \otimes \mathcal{U}$.

We now describe the basic construction which will be needed to get our Popa algebras. Let $\pi : E \to M_k(\mathbb{C})$ be a representation, $\tau \in \mathrm{T}(E)$ and $\epsilon > 0$. Choose $n \in \mathbb{N}$ very large and consider the map $\rho : E \to E \otimes M_n(\mathbb{C})$ given by

$$x \mapsto 1_E \otimes diag(0_{n-k}, \pi(x)) + x \otimes diag(1_{n-k}, 0_k),$$

where $diag(0_{n-k}, \pi(x))$ is the block diagonal element in $M_n(\mathbb{C})$ whose first $n - k$ entries down the diagonal are zero and the bottom block is given by $\pi(x)$, while $diag(1_{n-k}, 0_k) \in M_n(\mathbb{C})$ has $n - k$ 1's down the diagonal followed by k zeros. The key remarks about this choice of connecting map are:

(1) If $\frac{n-k}{n} > 1 - \epsilon$ then $|\tau \otimes \text{tr}_n(\rho(x)) - \tau(x)| < 2\epsilon \|x\|$ for all $x \in E$. That is, in trace the connecting map ρ is almost the same as the map $x \mapsto x \otimes 1_{M_n}$ (which would be the natural connecting maps to use if we were trying to construct $E \otimes \mathcal{U}$).
(2) If $I \subset E$ is an ideal and there exists an element $x \in I$ such that $\pi(x) \neq 0$ then the ideal generated by $\rho(I)$ is all of $E \otimes M_n(\mathbb{C})$. This follows from the definition of ρ and the simplicity of $M_n(\mathbb{C})$. It is this fact that will allow us to deduce simplicity of our inductive limits.
(3) There exists a finite dimensional C*-algebra $B \subset E \otimes M_n(\mathbb{C})$ with unit e such that $e\rho(x) - \rho(x)e = 0$ and $e\rho(x)e \in B$, for all $x \in E$. (Let $B = diag(0, \ldots, 0, \pi(E)) \subset M_n(\mathbb{C})$.) This remark will immediately imply that our inductive limits satisfy the finite dimensional approximation property which defines Popa algebras.
(4) The representation $\pi \otimes \text{id} : E \otimes M_n(\mathbb{C}) \to M_k \otimes M_n(\mathbb{C})$ is again a finite dimensional representation and hence this whole procedure can be reapplied to the algebra $E \otimes M_n(\mathbb{C})$ (thus yielding an inductive system).

We now enter the gory details. So let E be a residually finite dimensional C*-algebra and $\pi_i : E \to M_{k(i)}(\mathbb{C})$ be a separating sequence of representations. In fact, we will assume that for every $x \in E$, $\|x\| = \lim_i \|\pi_i(x)\|$ (taking direct sums, it is not hard to see that every residually finite dimensional C*-algebra has such a sequence). Note that for every $n \in \mathbb{N}$, $\pi_i \otimes id : E \otimes M_n(\mathbb{C}) \to M_{k(i)} \otimes M_n(\mathbb{C})$ is a separating sequence of the same type.

Now choose natural numbers $1 = n(0) \leq n(1) \leq n(2) \leq \ldots$ such that
$$\frac{n(0)n(1) \cdots n(j-1)k(j)}{n(j)} < 2^{-j},$$
for all $j \in \mathbb{N}$. One then defines algebras $E = E_0, E_1 = E_0 \otimes M_{n(1)}, E_2 = E_1 \otimes M_{n(2)}, E_3 = E_2 \otimes M_{n(3)}, \ldots$ and inclusions $\rho_i : E_i \hookrightarrow E_{i+1}$ as in the basic construction described above where the inclusion ρ_i uses the finite dimensional representation $\pi_{i+1} \otimes id \otimes \cdots \otimes id : E \otimes M_{n(1)} \otimes \cdots \otimes M_{n(i)} \to M_{k(i+1)} \otimes M_{n(1)} \otimes \cdots \otimes M_{n(i)}$ in the lower right hand corner. Letting $\Phi_{j,i} : E_i \to E_j$, $i \leq j$, be defined by $\Phi_{j,i} = \rho_{j-1} \circ \cdots \circ \rho_i$ we get an inductive system $\{E_i, \Phi_{j,i}\}$.

There are some projections in the above inductive system which we will need. Let $P_i \in E_i$ be the projection
$$P_i = 1_{E_{i-1}} \otimes diag(1_{n(i)-n(1)\cdots n(i-1)k(i)}, 0_{n(1)\cdots n(i-1)k(i)}).$$
Note that P_{i+1} commutes with all of $\rho_i(E_i)$ (and, in particular, with $\rho_i(P_i)$). Note also that if we write $E_{i+1} = E_i \otimes M_{n(i+1)}$ then
$$P_{i+1}\rho_i(P_i) = P_i \otimes diag(1_{n(i+1)-n(1)\cdots n(i)k(i+1)}, 0_{n(1)\cdots n(i)k(i+1)}).$$

Letting A be the inductive limit of the inductive system above, we only have to show that A is the Popa algebra we are after.

Proof of Theorem 5.3.1: We keep all the notation above. We leave it to the reader to verify that A is a Popa algebra as this follows from our remarks above and the construction of A. (That A is unital and satisfies the right finite dimensional approximation property is obvious while simplicity follows from the remark that any ideal in A must eventually intersect some E_i (cf. [**27**, Lemma III.4.1]).) Note also that A was constructed as an inductive limit of matrices over E and hence belongs to the class $\mathfrak{C}$ when E does.

5.3. II$_1$-FACTOR REPRESENTATIONS OF POPA ALGEBRAS

Now observe that given a trace $\tau \in T(E)$ we can define traces $\tau_j \in T(E_j)$ by $\tau_j = \tau \otimes \text{tr}_{n(1)} \otimes \cdots \otimes \text{tr}_{n(j)}$. Then the embedding $\rho_j : E_j \to E_{j+1}$ almost intertwines τ_j and τ_{j+1}. More precisely, a straightforward (but rather unpleasant) calculation shows that for $i < j$,

$$\tau_j(\Phi_{j,i}(x)) = \frac{\prod_{s=i}^{j-1}(n(s+1) - n(1)n(2)\cdots n(s)k(s+1))}{\prod_{s=i}^{j-1} n(s+1)} \tau_i(x) + \lambda_{i,j}\eta_{i,j}(x),$$

where $\lambda_{i,j} = 1 - \frac{\prod_{s=i}^{j-1}(n(s+1) - n(1)n(2)\cdots n(s)k(s+1))}{\prod_{s=i}^{j-1} n(s+1)}$ and $\eta_{i,j}$ is some tracial state on E_i.
Hence we get the estimate

$$|\tau_j(\Phi_{j,i}(x)) - \tau_i(x)| \leq 2\lambda_{i,j}\|x\|,$$

for all $x \in E_i$. But, it can be shown by induction that $\prod_{s=i}^{j-1}(1 - \frac{n(1)n(2)\cdots n(s)k(s+1)}{n(s+1)}) \geq \prod_{s=i}^{j-1}(1 - 2^{-s-1}) \geq 1 - 2^{-i} + 2^{-j} \geq 1 - 2^{-i}$, for all $i < j \in \mathbb{N}$. Hence we get that

$$|\lambda_{i,j}| = |1 - \prod_{s=i}^{j-1}(1 - \frac{n(1)n(2)\cdots n(s)k(s+1)}{n(s+1)})| \leq 2^{-i}.$$

We have almost established part (1) in Theorem 5.3.1. For each $i \in \mathbb{N}$, extend τ_i to a state on A (after identifying E_i with it's image in A). It is clear that if we take any weak-$*$ cluster point, γ, of this sequence then we will get a trace on A. Moreover, by the estimates above, we have that for each $x \in E_i$,

$$|\gamma(x) - \tau_i(x)| \leq 2^{-i}\|x\|.$$

Since we always have τ-preserving embeddings of E into E_i, it should be clear how to construct the embedding ρ in the statement of the theorem.

Our last task is to prove that $\pi_\gamma(A)'' \cong \pi_\tau(E)''\bar{\otimes}R$. To do this, we will need to study the projections $P_j \in E_j$ defined above. The idea is that we will use the P_j's to construct different projections $Q^{(i)} \in \pi_\gamma(A)''$ with the following properties:

(1) $Q^{(i)} = P_i Q^{(i+1)} = Q^{(i+1)} P_i$ and hence $Q^{(i)} \leq Q^{(i+1)}$ for all $i \in \mathbb{N}$.
(2) $Q^{(i+1)} \in \pi_\gamma(E_i)'$, for all $i \in \mathbb{N}$ (where we have identified E_i with it's image in A).
(3) $\gamma(Q^{(i)}) \geq 1 - 2^{-i}$.
(4) For each $i \in \mathbb{N}$, $\frac{\gamma(Q^{(i+1)}x)}{\gamma(Q^{(i+1)})} = \tau_i(x)$, for all $x \in E_i$.
(5) The natural inclusion of the weak closure of $Q^{(i)}\pi_\gamma(E_{i-1})Q^{(i)}$ into the weak closure of $Q^{(i+1)}\pi_\gamma(E_i)Q^{(i+1)}$ (which is a natural inclusion by (1) above) is isomorphic to the (non-unital) inclusion

$$\pi_{\tau_{i-1}}(E_{i-1})'' \hookrightarrow \pi_{\tau_i}(E_i)'' \cong \pi_{\tau_{i-1}}(E_{i-1})'' \otimes M_{n(i)}$$

given by

$$x \mapsto x \otimes diag(1_{n(i)-n(1)n(2)\cdots n(i-1)k(i)}, 0_{n(1)n(2)\cdots n(i-1)k(i)}).$$

We claim that the construction of such $Q^{(i)}$'s will complete the proof. Indeed, if we can do this then one uses part (5) and Elliott's approximate intertwining argument to compare the (non-unital) inclusions $Q^{(1)}\pi_\gamma(E_0) \subset Q^{(2)}\pi_\gamma(E_1) \subset \cdots$ to the natural (unital) inclusions $E_0 \subset E_0 \otimes M_{n(1)} \subset E_0 \otimes M_{n(1)} \otimes M_{n(2)} \subset \cdots$. Part (3) ensures that the former sequence recaptures $\pi_\gamma(A)''$ while the latter sequence gives $E_0 \otimes \mathcal{U}$ in the limit, where $\mathcal{U}$ is a UHF algebra, and hence the weak closure will be as desired and the proof will be complete.

The construction of the $Q^{(i)}$'s is fairly simple. For each $i \in \mathbb{N}$ we define projections $Q_n^{(i)} = \pi_\gamma(P_i P_{i+1} \cdots P_{i+n}) \in \pi_\gamma(A)''$. Since the $Q_n^{(i)}$'s are decreasing (as $n \to \infty$), there exists a strong operator topology limit. Define $Q^{(i)} = \text{sot}-\lim_{n\to\infty} Q_n^{(i)}$. Then $Q^{(i)}$ is a projection and it is straightforward to verify conditions (1) and (2) above. (Recall that P_j commutes with $\Phi_{j,i}(E_i)$ whenever $i < j$.) Thus we are left to verify the last three conditions.

Proof of (3). It suffices to show that $\gamma(Q_n^{(i)}) \geq 1 - 2^i - 2^{-i-n}$ for all n. But we may identify $Q_n^{(i)}$ with a projection in E_{i+n} and so using the first part of the proof of this theorem (and using the identification) we get $|\gamma(Q_n^{(i)}) - \tau_{i+n}(Q_n^{(i)})| < 2^{-i-n}$. However, it follows from the construction of $Q_n^{(i)}$ that

$$\tau_{i+n}(Q_n^{(i)}) = \prod_{s=1}^{n} \text{tr}_{n(i+s)}(diag(1_{n(i+s)-n(1)\cdots n(i+s-1)k(i+s)}, 0_{n(1)\cdots n(i+s-1)k(i+s)})).$$

Thus, by the calculations given in the first part of the proof of this theorem, we see that $\tau_{i+n}(Q_n^{(i)}) \geq 1 - 2^{-i}$ and hence $\gamma(Q_n^{(i)}) \geq 1 - 2^i - 2^{-i-n}$.

Proof of (4). We must show that if $x \in E_{i-1}$ then

$$\tau_{i-1}(x)\gamma(Q^{(i)}) = \lim_{n\to\infty} \gamma(Q_n^{(i)} x).$$

In order to show this, it suffices to prove that

$$\tau_{i-1}(x) = \frac{\tau_{i+n}(Q_n^{(i)} x)}{\tau_{i+n}(Q_n^{(i)})},$$

for all $n \in \mathbb{N}$. This last equality, however, is evident from the construction.

Proof of (5). It suffices to show (essentially due to the uniqueness, up to unitary equivalence, of GNS representations together with part (4)) that there exists a surjective $*$-homomorphism $\eta : E_{i-1} \otimes M_{n(i)} \to Q^{(i+1)}\pi_\gamma(E_i)$ such that for every $x \in E_{i-1}$,

$$\eta(x \otimes diag(1_{n(i)-n(1)\cdots n(i-1)k(i)}, 0_{n(1)\cdots n(i-1)k(i)})) = Q^{(i)}\pi_\gamma(x) = Q^{(i+1)}P_i \pi_\gamma(x).$$

But since

$$P_i \rho_{i-1}(x) = x \otimes diag(1_{n(i)-n(1)\cdots n(i-1)k(i)}, 0_{n(1)\cdots n(i-1)k(i)}) \in E_i = E_{i-1} \otimes M_{n(i)},$$

we get the desired homomorphism by identifying $E_{i-1} \otimes M_{n(i)}$ with it's image in A, passing to the GNS construction π_γ and then cutting down by $Q^{(i+1)}$. $\square$

With the technical preliminaries now out of the way we can address Popa's question concerning the set of II_1-factors arising from representations of Popa algebras. Our previous work on amenable traces proves relevant in at least one case.

THEOREM 5.3.2. *Assume A is a locally reflexive Popa algebra with unique trace τ. Then $\pi_\tau(A)'' \cong R$.*

PROOF. Since A is quasidiagonal τ must be a quasidiagonal trace. In particular, it must be an amenable trace and hence, by Corollary 4.3.4, produce the hyperfinite II_1-factor in the GNS construction. □

Note that we really didn't need A to be a Popa algebra in the previous result as this was only used in order to deduce quasidiagonality. Indeed, if A is any locally reflexive, quasidiagonal C*-algebra then A must have at least one trace whose GNS representation gives either a matrix algebra or R^5 and hence in the unique trace case we are forced to get something hyperfinite.

We will eventually see that the theorem above is false without the assumption of local reflexivity. However, we now wish to observe just how rich the II_1-factor representation theory of a Popa algebra can be in general. Recall that a II_1-factor is called McDuff if it is isomorphic to something of the form $N \bar{\otimes} R$. Also recall that such a factor is hyperfinite if and only if $N \cong R$ (and hence McDuff factors are almost never hyperfinite).

THEOREM 5.3.3. *There exists a Popa algebra A with the property that for each McDuff factor, M, there exists a trace $\tau_M \in T(A)$ such that $\pi_\tau(A)'' \cong M$.*

PROOF. Since every II_1-factor arises as the weak closure of a GNS representation of $C^*(\mathbb{F}_\infty)$, and $C^*(\mathbb{F}_\infty)$ is residually finite dimensional, this theorem follows immediately from Theorem 5.3.1. □

One cute consequence of this result is that every McDuff factor has some approximation properties on a dense subalgebra which are close to some of the various characterizations of R (though most McDuff factors are not hyperfinite). Compare with Theorem 6.3.3 to see the analogous statements which characterize R.

COROLLARY 5.3.4. *If $M \subset B(L^2(M))$ is a McDuff factor then there exists a weakly dense C*-subalgebra $A \subset M$ such that:*
(1) *There exist finite rank projections, $P_1 \leq P_2 \leq \ldots$, such that $\|[P_n, a]\| \to 0$ for all $a \in A$.*
(2) *There exists a state ϕ on $B(L^2(M))$ such that $A \subset B(L^2(M))_\phi = \{T \in B(L^2(M)) : \phi(TS) = \phi(ST), S \in B(L^2(M))\}$.*
(3) *There exists a sequence of II_1-factors, $R_n \subset B(L^2(M))$, such that $R_n \cong R$ for all n and for each $a \in A$ we can find $x_n \in R_n$ such that $\|a - x_n\| \to 0$.*
(4) *There exists a sequence of normal, u.c.p. maps $\varphi_n : M \to M_{k(n)}(\mathbb{C})$ such that $\|\varphi_n(ab) - \varphi_n(a)\varphi_n(b)\|_2 \to 0$ for all $a, b \in A$.*
(5) *There exists a completely positively liftable u.c.p. map $\Phi : M \to R^\omega$ such that $\Phi|_A$ is a *-monomorphism.*

PROOF. Let A be the universal Popa algebra from the previous theorem. With the exception of statement (3), all assertions above follow from the quasidiagonality of A (identified with a weakly dense subalgebra of M). Indeed, (1) follows from the definition of quasidiagonality plus Voiculescu's theorem (since a II_1-factor never contains any non-zero compact operators). To see property (4) we first note that such maps exist on A and hence we can extend them to u.c.p. maps on all of M. Of course, the extensions need not be normal but, as we have seen before, we can

---
[5]Use the fact that AT(A) is not empty, by quasidiagonality, and is a face in $\tilde{T}(A)$, hence contains an extreme point of T(A). It follows that A must have at least one factorial amenable trace and this will do thanks to Corollary 4.3.4.

always perturb to normal maps without destroying the asymptotic multiplicativity on A. Note that (5) follows from (4) via an argument similar to the one we saw in the proof of (1) $\implies$ (2) from Theorem 3.1.7. Finally, statement (2) is just asserting that A is contained in the centralizer of some state on $B(L^2(M))$ which is to say that A has at least one amenable trace. However, A has at least one quasidiagonal trace (actually it has tons of amenable traces) which is enough to show (2).

Hence we are left to demonstrate (3). However, this follows from Voiculescu's theorem since we can also find a faithful representation of A into R and this representation must be approximately unitarily equivalent to the inclusion $A \subset M \subset B(L^2(M))$. □

When we first discovered part (5) in the corollary above, we thought that it may be useful in showing that every II_1-factor embeds into R^ω. However, if the map Φ is normal – which would imply that it is a $*$-homomorphism since it is so on a weakly dense subalgebra – then $M \cong R$ and hence there is no hope of proving normality in general.

Of course, if a Popa algebra (or any other C*-algebra) is nuclear then all of its representations give hyperfinite von Neumann algebras. The next larger class of algebras is the class of exact C*-algebras and our work shows that as soon as one leaves the class where hyperfiniteness is automatic one finds counterexamples to Popa's question concerning the II_1-factor representation theory of Popa algebras.

THEOREM 5.3.5. *There exists an exact, Popa algebra, A, with non-hyperfinite II_1-factor representations.*

PROOF. According to the proof of Corollary 4.3.8 there is an exact residually finite dimensional C*-algebra B with a trace τ such that $\pi_\tau(B)''$ is a free group factor. Applying Theorem 5.3.1 to this example we get an exact Popa algebra with a trace whose GNS representation yields $\pi_\tau(B)'' \overline{\otimes} R$ (which is not hyperfinite). □

Another curious consequence of this work is that McDuff factors which are generated by exact C*-algebras always have 'norm microstates' on a dense subalgebra.

THEOREM 5.3.6. *If $M \subset B(L^2(M))$ is McDuff and contains a weakly dense, exact C*-subalgebra then there exists a weakly dense C*-subalgebra $A \subset M$ and finite dimensional matrix subalgebras $M_n \subset B(L^2(M))$ such that for each $a \in A$ there exists a sequence $a_n \in M_n$ such that $\|a - a_n\| \to 0$. (Hence, for every noncommutative polynomial P in k variables and finite set $\{a^{(i)}\}_{i=1}^k \subset A$ we have $\|P(a^{(1)}, \ldots, a^{(k)}) - P(a_n^{(1)}, \ldots, a_n^{(k)})_n)\| \to 0$ as $n \to \infty$.)*

PROOF. Since every exact C*-algebra is the quotient of an exact, residually finite dimensional C*-algebra (cf. [**17**, Corollary 5.3]), it follows from Theorem 5.3.1 that M contains a weakly dense, exact Popa algebra. In particular, M contains a weakly dense, exact, quasidiagonal C*-algebra. Since M is a factor, it can't contain any nonzero compact operators and hence the result now follows from [**25**] (see also [**18**] for the general case). □

Note that the preceding theorem covers many group von Neumann algebras (e.g. $\Gamma = G_1 \times G_2$ where G_1 is discrete, amenable and i.c.c. while G_2 is discrete, exact and i.c.c.). We are not, however, claiming that this result implies R^ω embeddability for such group von Neumann algebras. Indeed, it is not at all clear that the existence of norm microstates implies the existence of 'weak' microstates (in the sense of Voiculescu) since there does not appear to be any way of understanding how the traces behave on the norm approximations.

CHAPTER 6

Applications and connections with other areas

In the final chapter of these notes we wish to consider a number of applications and connections with problems which, on the surface, seem to be quite far removed from finite representation theory.

6.1. Elliott's classification program

The classification program has been an ambitious attempt to find complete isomorphism invariants for the class of nuclear C*-algebras. At this level of generality it is really impossible to imagine that such an invariant, if it exists, could ever be 'computable' in general. For example, there is a one-to-one correspondence between compact metric spaces and (unital, separable) abelian C*-algebras and hence such a classification would necessarily include a classification, up to homeomorphism, of all compact metric spaces. On the other hand, there are results in geometry which state, under suitable 'rigidity' assumptions, that manifolds which are only assumed homotopy equivalent must in fact be diffeomorphic. Also group theorists had a complete classification of the finite *simple* groups two decades ago but no such result is expected, as far as we know, for non-simple finite groups. Our point is that though a very general classification theorem for nuclear C*-algebras may exist in principle it is unlikely to be very applicable due to the inherit difficulty of computing whatever invariant is capable of completely classifying such a broad class of algebras. However, if we look to other fields of mathematics for inspiration, then it does not seem unreasonable to expect that if one further assumes some sort of 'rigidity', in addition to nuclearity, then classification results with computable invariants might be possible. Exactly what 'rigid' means in this context is not yet clear but most operator algebraists would agree that requiring *simplicity*, like in the case of finite groups, is a reasonable start.

The invariant which has been most successful, and hence attracted the most attention, is known as the Elliott invariant which we now describe.[1] For a nuclear C*-algebra A, the Elliott invariant is the triple $(K_0(A), K_1(A), \mathrm{T}(A))$, where $\mathrm{T}(A)$ is the set of tracial states on A, together with the natural pairing $P_A : K_0(A) \times \mathrm{T}(A) \to \mathbb{R}$.[2] Given two algebras A and B, we say that their Elliott invariants are isomorphic if $K_1(A) \cong K_1(B)$ and there exist a scaled, ordered group isomorphism $\Phi : K_0(A) \to K_0(B)$ and an affine homeomorphism $T : \mathrm{T}(A) \to \mathrm{T}(B)$ such that $P_A(x, \tau) = P_B(\Phi(x), T(\tau))$, for all $(x, \tau) \in K_0(A) \times \mathrm{T}(A)$.

---

[1] Many thanks to M. Rørdam and H. Lin for some helpful discussions regarding various issues discussed in this section.

[2] It is quite possible that the tracial state space be empty and then, of course, there is no pairing with $K_0(A)$.

For some time it was felt that the Elliott invariant may be a complete invariant for the class of *simple* nuclear C*-algebras. However examples of Rørdam definitively showed that this was not the case (cf. [**72**]). But, his examples were not stably finite and one could still hope that the Elliott invariant may be complete for simple, stably finite, nuclear C*-algebras. Unfortunately, some recent work of Toms (cf. [**75**]) has shown that this is also not possible as he has constructed examples of non-isomorphic simple ASH algebras with the same Elliott invariant.

On the other hand, some experts have suggested that the proper class of 'rigid' algebras to expect general classification results is for the simple, nuclear C*-algebras of real rank zero or stable rank one (or both). Since this still contains many of the 'naturally occurring' examples of finite simple nuclear C*-algebras (e.g. irrational rotation algebras, crossed products of the Cantor set by minimal homeomorphisms, the crossed product of the CAR algebra by the noncommutative Bernoulli shift and others) there is still good reason to pursue a general classification theorem in this setting.

If we restrict to the class of algebras with stable rank one (adding real rank zero later on) then one can formulate Elliott's conjecture as follows.

(Special case of) **Elliott's Conjecture:** If two simple, nuclear C*-algebras of stable rank one have isomorphic Elliott invariants (as described above) then they are isomorphic.

To the untrained eye this special case may look very special indeed. However, it is still unimaginably general and it seems to the present author that we are a long way from the resolution of this 'special' case of Elliott's conjecture. However, it is our hope that if we add a few more hypotheses, such as real rank zero, quasidiagonality, weakly unperforated invariant[3] and unique trace then classification results may now be in sight. Moreover, our feeling is that the key to unlocking Elliott's conjecture (at least in the real rank zero, stable rank one case) is approximation properties of traces.

As evidence to support this point of view we now (a) present a number of classification results where uniform locally finite dimensional traces play an essential role, (b) point out which tracial approximation question needs to be resolved in order to complete the general unique trace case mentioned above and (c) recall some facts which indicate there is reasonable hope of carrying out part (b). Finally we will end this section with a number of predictions of Elliott's conjecture as they may be of independent interest.

So let's have a look at some classification results where uniform locally finite dimensional traces play an important role. As discussed in the introduction to this paper our feeling is that Huaxin Lin's classification theorem for tracially AF algebras is the 'right' classification theorem (in the real rank zero, stable rank one, weakly unperforated case) and hence we have no intention of presenting more general classification theorems. Rather our goal will be to show that certain examples are already covered by Lin's theorem and the way to do this is to use tracial approximation properties to show that these examples are in fact tracially AF. We will begin with some 'rational' classification results. It will be convenient to summarize (special cases of) some known facts.

---

[3]See the section on tracially AF algebras for the definitions of these things.

THEOREM 6.1.1 (cf. [**70**], [**71**], [**12**]). [4] *Assume A is simple, nuclear and has a unique tracial state. Then for any UHF algebra $\mathcal{U}$ the (simple, nuclear) tensor product algebra $A \otimes \mathcal{U}$ has a unique trace, real rank zero, stable rank one, Blackadar's fundamental comparison property*[5] *and the Riesz property.*[6]

THEOREM 6.1.2. *If A is simple, nuclear and has a unique trace τ which happens to be uniform locally finite dimensional then for any UHF algebra $\mathcal{U}$ we have that $A \otimes \mathcal{U}$ is tracially AF.*

PROOF. Since $A \otimes \mathcal{U}$ has real rank zero, stable rank one and weakly unperforated K-theory this result follows from our tracial characterization of tracially AF algebras (cf. Proposition 4.5.5) together with the fact that the unique trace on $A \otimes \mathcal{U}$ is also uniform locally finite dimensional by Proposition 3.5.7. □

COROLLARY 6.1.3. [7] *Let M be a compact manifold and $h : M \to M$ a minimal diffeomorphism. If $C(M) \rtimes_h \mathbb{Z}$ has a unique trace then for each UHF algebra $\mathcal{U}$ it follows that $(C(M) \rtimes_h \mathbb{Z}) \otimes \mathcal{U}$ is isomorphic to one of the AH algebras considered in* [**30**].

PROOF. Letting $A = C(M) \rtimes_h \mathbb{Z}$ in the previous theorem, simplicity and nuclearity are well known and we have assumed a unique tracial state. Hence we only have to know that A satisfies the Universal Coefficient Theorem (UCT) of Rosenberg-Schochet – which it does by results of Rosenberg-Schochet [**73**] – and that the unique trace is uniform locally finite dimensional for then the previous theorem will imply that $(C(M) \rtimes_h \mathbb{Z}) \otimes \mathcal{U}$ is a nuclear, tracially AF algebra satisfying the UCT and hence it must be isomorphic to one of the AH algebras considered in [**30**] by Lin's classification theorem [**50**]. However, the unique trace is locally finite dimensional since the remarkable work of Q. Lin and Phillips (cf. [**54**]) implies that $C(M) \rtimes_h \mathbb{Z}$ is an inductive limit of type I C*-algebras (actually very special 'recursive' subhomogeneous algebras) and hence Corollary 4.4.4 applies. □

Note that a similar result can be formulated for simple inductive limits of type I algebras which have unique trace (i.e. all of these become AH algebras after tensoring with a UHF algebra).

Our assumption of a unique trace above was essential to ensure that the rationalization had real rank zero. Since a unique trace does not imply real rank zero in general we will have to add this assumption (along with stable rank one and weak unperforation which also need not be automatic) if we wish to drop the rationalization procedure. However, once these assumptions are added then one can relax the unique trace hypothesis to allow a norm separable tracial space as in Proposition 4.5.5 and still obtain classification results. But since many people seem to find the unique trace formulation more aesthetically pleasing we will continue to use this assumption (though it can be relaxed in all of the classification results below).

---

[4] If one assumes that A is also approximately divisible in the sense of [**12**] then one need not tensor with the UHF algebra in this theorem.

[5] This means that if $p, q \in A$ are projections and $\tau(p) < \tau(q)$ for all $\tau \in \mathrm{T}(A)$ then p is Murray-von Neumann equivalent to a subprojection of q. Note that this implies weak unperforation.

[6] In other words, if x, y_1, y_2 are positive elements of $K_0(A)$ and $x \leq y_1 + y_2$ then there exist $z_1, z_2 \in K_0(A)$ such that $x = z_1 + z_2$ and $z_i \leq y_i$, $i = 1, 2$.

[7] Though we haven't seen it written down anywhere, this corollary may be known to the experts as it can also be deduced from the results of [**53**], for example.

THEOREM 6.1.4. *Let A be a simple, nuclear C^*-algebra with real rank zero, stable rank one, weakly unperforated invariant and satisfying the UCT ([**73**]). If A has a unique trace τ which happens to be uniform locally finite dimensional then A is isomorphic to one of the AH algebras considered in [**30**]).*

PROOF. According to Proposition 4.5.5 we don't even need to know nuclearity or the UCT in order to deduce that A is tracially AF. However, these two assumptions are certainly needed to apply Lin's classification theorem [**50**]. □

DEFINITION 6.1.5. Given $\tau \in \mathrm{T}(A)$, we will say that A is τ-*tracially type I* if for each finite subset $\mathfrak{F} \subset A$ and $\varepsilon > 0$ there exists a type I subalgebra $B \subset A$ with unit e such that $\|[x,e]\| < \varepsilon$ for all $x \in \mathfrak{F}$, $e\mathfrak{F}e \subset^{\varepsilon} B$ and $\tau(e) > 1 - \varepsilon$.

The definition above is admittedly artificial. However, it allows us to treat both inductive limits of type I C*-algebras (even "locally type I" algebras) and C*-algebras with finite tracial topological rank (in the sense of either [**52**, Definition 3.1] or [**52**, Definition 3.4]) at the same time as both of these classes are easily seen to be τ-tracially type I with respect to every $\tau \in \mathrm{T}(A)$.

COROLLARY 6.1.6. *Assume that A is nuclear, has unique trace τ, is τ-tracially type I, simple, real rank zero, stable rank one, weakly unperforated K-theory and satisfies the UCT. Then A is isomorphic to an AH-algebra.*

PROOF. By the previous theorem, we only need to show that τ is a uniform locally finite dimensional trace. The proof of this is similar to the proof of Proposition 4.5.5 but also requires Lemma 4.4.1 and Lemma 4.4.3.

Indeed, according to Lemma 4.4.1 it suffices to show that for each finite set $\mathfrak{F} \subset A$ and $\epsilon > 0$ there exists a C*-subalgebra $C \subset A$ and a u.c.p. map $\phi : C \to M_n(\mathbb{C})$ such that $d(\mathfrak{F}, C_\phi) < \epsilon$ and $\|\tau|_C - \mathrm{tr}_n \circ \phi\|_{C^*} < \epsilon$. But, by definition of τ-tracially type I we can find a type I subalgebra $B \subset A$ with unit e satisfying the three conditions stated above. If we let $C = e^\perp A e^\perp \oplus B$ then C nearly contains $\mathfrak{F}$ in norm and if we knew that the trace $\tau|_C$ was uniform locally finite dimensional then we would be done. While that may or may not be true, it is a fact that $\tau|_C$ is very close in norm to a uniform locally finite dimensional trace (and this is good enough to complete the proof). Indeed, since B is a type I algebra the tracial state on C defined by composing the quotient map $C \to B$ with $\frac{1}{\tau(e)}\tau(\cdot)$ yields a type I von Neumann algebra in its GNS representation and hence, by Lemma 4.4.3, is uniform locally finite dimensional. But this trace is also close in norm to $\tau|_C$ (since $\tau(e) > 1 - \epsilon$) and hence the proof is complete. □

REMARK 6.1.7. The point of the previous corollary is that basically anything which is built out of type I C*-algebras (either via classical constructions like inductive limits or Lin's more recent "measurable" approximations) satisfies the equation $\mathrm{T}(A) = \mathrm{UAT}(A)_{\mathrm{LFD}}$. Hence in the presence of nuclearity, simplicity, real rank zero, stable rank one, weak unperforation and a natural separability condition on the size of the tracial space one immediately gets classification results for such algebras.

We should also mention that if A has real rank zero and finite decomposition rank in the sense of Kirchberg-Winter [**48**] then it can be shown that $\mathrm{T}(A) = \mathrm{UAT}(A)_{\mathrm{LFD}}$. Hence this class of algebras also fits into the tracial approximation picture. However, for the purposes of classification there is currently little reason to take this point of view as Wilhelm Winter has shown that the structural assumptions involved in finite decomposition rank allow one to prove much nicer classification theorems than we can presently achieve only from assumptions on traces (cf. [**82**]).

The next three results were first proved by Huaxin Lin, but they are also simple consequences of Corollary 6.1.6. Note that in all of these corollaries the role played by uniform locally finite dimensional traces appears to be crucial. Indeed we are not aware of an argument, other than the tracial approximation approach, which will provide the large finite dimensional algebras required to deduce tracially AF in any of the cases considered below.

COROLLARY 6.1.8. [**53**, Theorem 5.16] *Every simple, real rank zero, stable rank one, weakly unperforated, inductive limit of type I C^*-algebras with unique trace is isomorphic to an AH-algebra as in* [**30**].

COROLLARY 6.1.9. [**52**, Theorem 7.7] *Every simple, nuclear, C^*-algebra with finite tracial topological rank, satisfying the UCT and having a unique trace is isomorphic to an AH-algebra as in* [**30**].

PROOF. In [**52**] Lin shows that these hypotheses imply real rank zero, stable rank one and weakly unperforated K-theory. □

COROLLARY 6.1.10. [**53**, Corollary 5.5] *Let $h : M \to M$ be a minimal diffeomorphism of a compact manifold and $C(X) \rtimes_h \mathbb{Z}$ the corresponding crossed product. If $C(X) \rtimes_h \mathbb{Z}$ has a unique trace τ and $\tau_*(K_0(C(X) \rtimes_h \mathbb{Z}))$ is dense in $\mathbb{R}$ then it is isomorphic to an AH-algebra as in* [**30**].

PROOF. By [**54**], such crossed products are inductive limits of 'recursive subhomogeneous' algebras (which, in particular, are type I), they have stable rank one and weakly unperforated K-theory by [**62**] and, finally, real rank zero by [**63**]. □

REMARK 6.1.11. Note that the first two results above hold true under the weaker assumption of norm separability of the tracial space. However, Corollary 6.1.10 does not generalize quite so easily. As proved by Phillips in [**63**], if one knows that the canonical map $K_0(C(X) \rtimes_\phi \mathbb{Z}) \to Aff(\mathrm{T}(C(X) \rtimes_\phi \mathbb{Z}))$ has dense range then real rank zero follows and then, together with norm separability of the tracial space, one deduces that $C(X) \rtimes_\phi \mathbb{Z}$ is isomorphic to an AH algebra.

We now discuss what needs to be done in order to complete the case of Elliott's conjecture mentioned at the beginning of this section as well as point out why there is reason to hope that this may be possible.

COROLLARY 6.1.12. *Let A be a simple, nuclear C^*-algebra with real rank zero, stable rank one, weakly unperforated invariant, unique trace τ and satisfying the UCT. Then the following are equivalent:*

(1) *A is isomorphic to one of the AH algebras considered in* [**30**].
(2) *A is an inductive limit of type I subalgebras.*
(3) *A is an inductive limit of subalgebras A_i such that $\pi_\tau(A_i)''$ is a type I von Neumann algebra for each $i \in \mathbb{N}$.*
(4) *τ is uniform locally finite dimensional.*

PROOF. Evidently the last statement is the weakest and Theorem 6.1.4 says that (4) $\Longrightarrow$ (1) so we are done. □

Of course, the corollary above is just a reformulation of things we have already seen but we are trying to drive home the point that *the real rank zero, stable rank one, weakly unperforated, UCT, unique trace case of Elliott's conjecture is equivalent to proving that, under the same assumptions, the unique trace is always a uniform locally finite dimensional trace*. Now, we don't mean to give the impression that verifying this tracial approximation property will be trivial. However, the problem has now been isolated and the next two results show that if *either* of the conditions defining uniform locally finite dimensional traces are relaxed – and the assumption of quasidiagonality is added – then the resulting approximation property does, in fact, always hold. In other words, we believe there is evidence in favor of using uniform locally finite dimensional traces to complete the case of Elliott's conjecture described above.

THEOREM 6.1.13 (Special case of Theorem 4.3.3). *Let A be a nuclear, quasidiagonal C^*-algebra with unique trace τ. Then τ is a uniform quasidiagonal trace.*

THEOREM 6.1.14. *Let A be a simple, quasidiagonal C^*-algebra with unique trace τ. Then τ is locally finite dimensional.*

PROOF. Assume $A \subset B(H)$ is a faithful, irreducible representation (which exists by simplicity). Since A is quasidiagonal we can find finite rank projections $P_1 \leq P_2 \leq \cdots$ converging strongly to the identity and such that $\|[a, P_n]\| \to 0$ for all $a \in A$ (note that A is necessarily in general position – i.e. contains no compacts – since A is simple and unital). Define u.c.p. maps $\phi_n : A \to P_n B(H) P_n$ by $\phi_n(a) = P_n a P_n$ and note that, by uniqueness of τ, $\text{tr}_{rank(P_n)} \circ \phi_n(a) \to \tau(a)$ for all $a \in A$. Since the representation is irreducible, it can be shown (cf. [**14**, Corollary 3.5]) that for every $a \in A$, the distance from a to the multiplicative domain of ϕ_n tends to zero and hence τ is locally finite dimensional. □

COROLLARY 6.1.15. *If A is simple, nuclear, quasidiagonal and has unique trace τ then $\tau \in \text{UAT(A)}_{\text{QD}} \cap \text{AT(A)}_{\text{LFD}}$.*

Note that Theorem 6.1.14 has absolutely nothing to do with nuclearity. Thus a general strategy naturally presents itself; take the maps provided by Theorem 6.1.14 and use nuclearity (or, perhaps, just local reflexivity as in the proof of Theorem 4.3.3?) to average them in a way which preserves multiplicative domains while forcing norm convergence to the trace τ. We remark, however, that nuclearity (or local reflexivity) *must* play a significant role in the proof as we will show in the next section that there exists a (non-locally reflexive) Popa algebra with unique trace which is not *uniform* locally finite dimensional (though it is, by Theorem 6.1.14, locally finite dimensional).

Finally, we should mention that Popa's work together with Theorem 4.3.3 provides another approach to attacking this problem. Indeed, we have the following result which is obviously relevant to this discussion as it shows that one can always find a finite family of finite dimensional subalgebras (as opposed to a single finite dimensional subalgebra as is required for tracially AF) such that a linear combination of their units is large in trace.

THEOREM 6.1.16. *Let A be a locally reflexive Popa algebra with real rank zero and unique trace τ. Then for every finite set $\mathcal{F} \subset A$ and $\varepsilon > 0$ there exists $n \in \mathbb{N}$, subalgebras $Q_1, \ldots, Q_m \subset A$ each of which is isomorphic to $M_n(\mathbb{C})$ and with units*

$e_1, \ldots, e_m$ such that $\|[e_i, x]\| < \varepsilon$ for $1 \leq i \leq m$ and all $x \in \mathcal{F}$, $e_i \mathcal{F} e_i \subset^\varepsilon Q_i$ for $1 \leq i \leq m$ and, finally,
$$\|\frac{1}{n}\sum_{k=1}^m e_k - 1_A\|_{\tau,1} < \varepsilon,$$
where $\|x\|_{\tau,1} = \tau(|x|)$.

PROOF. By [**67**, Theorem 3.3] it suffices to show that $\tau \in \mathrm{UAT}(A)_{\mathrm{QD}}$. However, as above this follows from Theorem 4.3.3 and hence we are done. □

We now want to spend some time looking at predictions of Elliott's conjecture – i.e. necessary conditions. It is quite interesting that various statements about quasidiagonality and approximation of traces naturally appear as necessary conditions for Elliott's conjecture to hold. We begin with a few simple facts which are known to the classification experts.

PROPOSITION 6.1.17. *Elliott's conjecture predicts that if A is a stably finite, simple, nuclear C^*-algebra and $\mathcal{M}_{2^\infty}$ denotes the CAR algebra then $A \otimes \mathcal{M}_{2^\infty}$ is an inductive limit of subhomogeneous algebras (i.e. ASH).*

PROOF. To deduce stable rank one and Blackadar's comparison property for the algebra $A \otimes \mathcal{M}_{2^\infty}$ we really only need to assume stable finiteness of A in Theorem 6.1.1. Hence $A \otimes \mathcal{M}_{2^\infty}$ is a simple, nuclear C*-algebra with stable rank one and weakly unperforated K-theory. But Elliott has shown that every possible weakly unperforated Elliott invariant arises from an ASH algebra (see, for example, the appendix of [**31**]) and hence we can find a simple ASH algebra B whose Elliott invariant is isomorphic to that of $A \otimes \mathcal{M}_{2^\infty}$. If B does not have stable rank one then we can replace B with $B \otimes \mathcal{M}_{2^\infty}$ to get a simple ASH algebra with stable rank one and Elliott invariant isomorphic to that of $A \otimes \mathcal{M}_{2^\infty}$ (since $A \otimes \mathcal{M}_{2^\infty} \otimes \mathcal{M}_{2^\infty} \cong A \otimes \mathcal{M}_{2^\infty}$). Thus our particular formulation of Elliott's conjecture indeed predicts that $A \otimes \mathcal{M}_{2^\infty}$ is ASH. □

COROLLARY 6.1.18. *Elliott's conjecture predicts that every simple, stably finite, nuclear C^*-algebra is quasidiagonal.*

PROOF. $A \otimes \mathcal{M}_{2^\infty}$ is quasidiagonal since it is an inductive limit of subhomogeneous algebras and quasidiagonality passes to subalgebras (i.e. $A \cong A \otimes 1 \subset A \otimes \mathcal{M}_{2^\infty}$). □

PROPOSITION 6.1.19. *Elliott's conjecture predicts that if A is simple, nuclear, stable rank one, real rank zero and has weakly unperforated invariant then A is tracially AF.*

PROOF. We first remark that in the real rank zero case the tracial simplex is no longer relevant and hence Elliott's invariant reduces to K-theory alone. Indeed, if both A and B are C*-algebras of real rank zero and $\Phi : K_0(A) \to K_0(B)$ is a scaled, ordered group isomorphism such that $\Phi([1_A]) = [1_B]$ then Φ induces an affine homeomorphism $\mathrm{T}(A) \to \mathrm{T}(B)$ since we may (affinely, homeomorphically) identify $\mathrm{T}(A)$ (resp. $\mathrm{T}(B)$) with the states in $\mathrm{Hom}(K_0(A), \mathbb{R})$ (resp. $\mathrm{Hom}(K_0(B), \mathbb{R})$). To see that this is true, we first note that the obvious map $\mathrm{T}(A) \to \mathrm{Hom}(K_0(A), \mathbb{R})$ is affine and injective since A has real rank zero. It is also onto the states in $\mathrm{Hom}(K_0(A), \mathbb{R})$ since every state on $K_0(A)$ comes from a trace on A when A is unital and exact (cf. [**36**]). Finally, it is easy to check (again using real rank zero) that a sequence of traces $\tau_n \in \mathrm{T}(A)$ converges to $\tau \in \mathrm{T}(A)$ in the weak-$*$ topology if and only if their images in $\mathrm{Hom}(K_0(A), \mathbb{R})$ converge in the topology of pointwise convergence and hence our identification is also a homeomorphism.

In [**30**] it is shown how to construct simple AH algebras with real rank zero and with arbitrary unperforated K-theory and Riesz interpolation property. Our assumptions on A also imply the Riesz property (cf. [**29**]) and hence Elliott's conjecture predicts that A is isomorphic to one of the AH algebras constructed in [**30**]. However, as observed by Lin [**51**, Proposition 2.6], the Elliott-Gong construction always yields tracially AF algebras and the proof is complete. □

Our next goal is to show that Elliott's conjecture predicts that every trace on every (not necessarily simple) nuclear, quasidiagonal C*-algebra is uniform quasidiagonal. (Recall that the corresponding statement for exact, quasidiagonal C*-algebras is not true – see Corollary 4.3.8.) The passage from the simple to non-simple cases is provided by the next lemma.

LEMMA 6.1.20. *If $\mathfrak{C}$ is a collection of C^*-algebras which contains $\mathbb{C}$ and is closed under i) increasing unions (i.e. inductive limits with injective connecting maps), ii) quasidiagonal, semi-split extensions (i.e. if $0 \to I \to E \to B \to 0$ is a semi-split (cf. [**11**]), short exact sequence, I contains an approximate unit of projections which is quasicentral in E and both $I, B \in \mathfrak{C}$ then $E \in \mathfrak{C}$) and iii) tensoring with finite dimensional matrix algebras then the following are equivalent:*

(1) $\mathrm{T}(A) = \mathrm{AT}(A)_{\mathrm{QD}}$ *(resp. $\mathrm{T}(A) = \mathrm{UAT}(A)_{\mathrm{QD}}$) for every quasidiagonal $A \in \mathfrak{C}$.*
(2) $\mathrm{T}(A) = \mathrm{AT}(A)_{\mathrm{QD}}$ *(resp. $\mathrm{T}(A) = \mathrm{UAT}(A)_{\mathrm{QD}}$) for every residually finite dimensional $A \in \mathfrak{C}$.*
(3) $\mathrm{T}(A) = \mathrm{AT}(A)_{\mathrm{QD}}$ *(resp. $\mathrm{T}(A) = \mathrm{UAT}(A)_{\mathrm{QD}}$) for every Popa algebra $A \in \mathfrak{C}$.*

If, moreover, the class $\mathfrak{C}$ is closed under tensor products with (non-unital) abelian algebras (it actually suffices to know $A \in \mathfrak{C} \implies A \otimes C_0((0,1]) \in \mathfrak{C}$) then the following are equivalent:

(4) $\mathrm{T}(A) = \mathrm{AT}(A)$ *(resp. $\mathrm{T}(A) = \mathrm{UAT}(A)$) for every $A \in \mathfrak{C}$.*
(5) $\mathrm{T}(A) = \mathrm{AT}(A)$ *(resp. $\mathrm{T}(A) = \mathrm{UAT}(A)$) for every quasidiagonal $A \in \mathfrak{C}$.*
(6) $\mathrm{T}(A) = \mathrm{AT}(A)$ *(resp. $\mathrm{T}(A) = \mathrm{UAT}(A)$) for every residually finite dimensional $A \in \mathfrak{C}$.*
(7) $\mathrm{T}(A) = \mathrm{AT}(A)$ *(resp. $\mathrm{T}(A) = \mathrm{UAT}(A)$) for every Popa algebra $A \in \mathfrak{C}$.*

PROOF. We first prove the equivalence of (1) - (3) and then indicate the changes necessary to prove the second part. The proofs are the same whether dealing with $\mathrm{AT}(A)_{\mathrm{QD}}$ or $\mathrm{UAT}(A)_{\mathrm{QD}}$ and hence we just treat the uniform quasidiagonal case.

(1) $\implies$ (3) is immediate. (3) $\implies$ (2) follows from Theorem 5.3.1.

(2) $\implies$ (1). Let $A \in \mathfrak{C}$, $\tau \in \mathrm{T}(A)$ and $\mathfrak{F} \subset A$ be an arbitrary finite set. Since A is quasidiagonal we can find a sequence of u.c.p. maps $\varphi_n : A \to M_{k(n)}(\mathbb{C})$ which are asymptotically multiplicative and asymptotically isometric (i.e. $\|a\| = \lim \|\varphi_n(a)\|$, for all $a \in A$). Passing to a subsequence, if necessary, we may assume that φ_1 (and all the other φ_n's) is as close to multiplicative on $\mathfrak{F}$ as we like. Put $\Phi = \oplus_n \varphi_n : A \to \Pi_n M_{k(n)}(\mathbb{C})$, and let E be the C*-algebra generated by $\Phi(A)$. Note that $\Phi : A \to E$ is as close to multiplicative on $\mathfrak{F}$ as we like, by construction.

Now observe that we have a semi-split, quasidiagonal, short exact sequence:

$$0 \to \oplus M_{k(n)}(\mathbb{C}) \to E + \oplus M_{k(n)}(\mathbb{C}) \to A \to 0.$$

Since $\mathfrak{C}$ is closed under all of the operations used, it follows that $E + \oplus M_{k(n)}(\mathbb{C}) \in \mathfrak{C}$ and it is clear that $E + \oplus M_{k(n)}(\mathbb{C})$ is residually finite dimensional. Hence every trace on $E + \oplus M_{k(n)}(\mathbb{C})$ is uniform quasidiagonal by (2). In particular, the trace $\tau \in T(A)$ induces such a trace on $E + \oplus M_{k(n)}(\mathbb{C})$ by composing with the quotient map. Since the splitting $\Phi : A \to E \subset E + \oplus M_{k(n)}(\mathbb{C})$ is almost multiplicative, it follows that we can construct a u.c.p. map on A (by composing maps on E with Φ) which is almost multiplicative on $\mathfrak{F}$ and which approximately recaptures the trace τ. This completes the proof of (2) $\implies$ (1).

For the equivalence of (4) - (7), we really only need to show the implication (5) $\implies$ (4) as the arguments above go through without change for the other implications. To prove this we will need Proposition 3.5.8 which says that amenable traces pass to quotients whenever there is a c.p. splitting for the quotient map and that uniform amenable traces always pass to quotients.

With those results in hand, (5) $\implies$ (4) becomes very simple. Indeed, let B be the unitization of the cone over A (i.e. the unitization of $C_0((0,1]) \otimes A$). Then B is quasidiagonal (cf. [77]) and belongs to $\mathfrak{C}$. Moreover, there is a natural surjective $*$-homomorphism $B \to A \oplus \mathbb{C} \to A$. A (non-unital) c.p. splitting for this quotient map is given by $a \mapsto e \otimes a$, where $e \in C_0((0,1])$ is any non-negative function such that $e(1) = 1$. Hence if every trace on B is (uniform) amenable then every trace on A enjoys the same property. $\square$

The assumptions on the class $\mathfrak{C}$ may seem unusual, but note that any one of the following classes of C*-algebras is closed under the operations needed in the lemma above: nuclear C*-algebras, exact C*-algebras (cf. [46, Section 7]), real rank zero C*-algebras (cf. [16, 2.10, 3.1, 3.14], it is easy to prove that if an extension is semi-split and quasidiagonal then every projection in the quotient lifts to a projection in the middle algebra) - though these algebras are not closed under tensoring with $C_0((0,1])$.

PROPOSITION 6.1.21. *If Elliott's conjecture holds for all nuclear, simple, quasidiagonal C*-algebras with stable rank one and weakly unperforated K-theory then $T(A) = UAT(A)_{QD}$ for every nuclear, quasidiagonal C*-algebra A.*

PROOF. We will apply the previous lemma to the set $\mathfrak{C}$ of all nuclear C*-algebras. We remark that extensions of nuclear C*-algebras are again nuclear by [20, Corollary 3.3].

So assume that Elliott's conjecture holds for all nuclear, simple, quasidiagonal C*-algebras with stable rank one and weakly unperforated K-theory and let A be quasidiagonal and nuclear. By the previous lemma, we may assume that A is simple. Let $\mathcal{M}_{2^\infty}$ be the CAR algebra and note that every trace on A extends (in fact, uniquely) to a trace on $A \otimes \mathcal{M}_{2^\infty}$. Hence it suffices to show that $T(A \otimes \mathcal{M}_{2^\infty}) = UAT(A \otimes \mathcal{M}_{2^\infty})_{LFD}$. (It is not clear that this will imply $T(A) = UAT(A)_{LFD}$, but it will certainly imply that $T(A) = UAT(A)_{QD}$.)

But just as in the proof of Proposition 6.1.17, it follows that $A \otimes B$ is an ASH algebra. In particular, it is an inductive limit of type I algebras and so from Corollary 4.4.4 we deduce that $T(A \otimes B) = UAT(A \otimes B)_{LFD}$. $\square$

Note that if every stably finite nuclear C*-algebra turns out to be quasidiagonal (recall that this question was asked by Blackadar and Kirchberg [13]) then Elliott's conjecture would predict that every trace on every nuclear C*-algebra is uniform quasidiagonal. Whether or not this happens in the non-simple case, we still have the following corollary.

COROLLARY 6.1.22. *If Elliott's conjecture holds for all nuclear, simple, quasidiagonal C*-algebras with stable rank one and unperforated K-theory then* $T(A) = UAT(A)_{QD}$ *for every simple, nuclear C*-algebra A.*

PROOF. If A is simple and not stably finite then A has no tracial states. On the other hand, if A is simple and stably finite then Elliott's conjecture predicts that A is also quasidiagonal. □

Note that the corollary above begs the following question: If A is nuclear and $\tau \in T(A)$ does there exist a *simple*, nuclear C*-algebra B with trace $\gamma \in T(B)$ and a *-homomorphism $\pi : A \to B$ such that $\tau = \gamma \circ \pi$? If this is the case then Elliott's conjecture predicts that every trace on every nuclear C*-algebra is uniform quasidiagonal.

We hope that the reader is now convinced that tracial approximation properties arise naturally as both necessary conditions (cf. Proposition 6.1.21) and sufficient conditions (cf. Theorem 6.1.4) for various cases of Elliott's conjecture. We wish to end this section with a possible approach to the general real rank zero, stable rank one, weakly unperforated case of Elliott's conjecture. Of course our focus is on tracial approximation so the following strategy is by no means the unique path to resolving this case of the conjecture.

Consider the following problems for a simple, nuclear, C*-algebra A with real rank zero, stable rank one, weakly unperforated K-theory and satisfying the UCT.

(1) Is it true that $AT(A)_{QD} \neq \emptyset$? (Recall that $T(A) = UAT(A) \neq \emptyset$ by nuclearity and stable finiteness.)
(2) If A is quasidiagonal then do we have $AT(A) = AT(A)_{QD} = AT(A)_{LFD}$? (Recall that this is true in the unique trace case.)
(3) Is it true that $UAT(A)_{QD} \cap AT(A)_{LFD} = UAT(A)_{LFD}$?
(4) Is it possible to remove the hypothesis of $\|\cdot\|_{A^*}$-separability from Proposition 4.5.5 in this setting?

PROPOSITION 6.1.23. *The real rank zero, stable rank one, weakly unperforated, UCT case of Elliott's conjecture is equivalent to affirmative answers to all four of the problems posed above.*

PROOF. Assume first that Elliott's conjecture holds in the case described above. It follows that if A is any C*-algebra with all of those properties then it must be tracially AF (cf. Proposition 6.1.19). Since $T(A) = UAT(A)_{LFD}$ for every tracially AF algebra it would follow that all four questions above must have affirmative answers.

Now suppose that one could prove all four questions above and we will show that it would follow that every A as above must necessarily be tracially AF. By problem (1) we would have that A is necessarily quasidiagonal since simplicity implies that any trace is faithful and we already saw in the proof of Proposition 4.1.3 that the existence of a faithful quasidiagonal trace implies that the C*-algebra

is quasidiagonal. If we knew that (2) also held then we would have that
$$T(A) = UAT(A)_{QD} = AT(A)_{LFD}$$
since nuclearity (trivially) implies $T(A) = AT(A)$ and (non-trivially) $AT(A)_{QD} = UAT(A)_{QD}$. In particular, problem (3) would then tell us that
$$T(A) = UAT(A)_{QD} \cap AT(A)_{LFD} = UAT(A)_{LFD}$$
and hence one would only have to remove the norm separability hypothesis from Proposition 4.5.5 (i.e. prove problem (4)) in order to deduce that A was tracially AF. Of course, Lin's classification theorem would then imply this case of Elliott's conjecture. □

6.2. Counterexamples to questions of Lin and Popa

Now that Huaxin Lin has essentially completed the tracially AF case of Elliott's conjecture (cf. [50]) it is clear that having abstract hypotheses which imply that a given C*-algebra is tracially AF is of fundamental importance. One result along these lines is Proposition 4.5.5. On page 694 of [51] Lin wrote "It is certainly tempting to conjecture that every quasidiagonal simple C*-algebra of real rank zero, stable rank one and with weakly unperforated K_0 is TAF (tracially AF)." Sorin Popa has asked (private communication) whether every Popa algebra with real rank zero and unique trace is tracially AF. The purpose of this section is to resolve these questions negatively. Essentially we show that no amount of K-theoretic assumptions alone will allow one to deduce that a Popa algebra is tracially AF. C*-algebraic structure (e.g. nuclearity or, perhaps, local reflexivity) must also be assumed. As we have tried to argue in the previous section, we believe that approximation properties of traces provide a reasonable strategy for proving that certain C*-algebras are tracially AF. In this section we will see that our study of finite representation theory also shows why certain algebras can't be tracially AF.

Before constructing our counterexamples we need a few remarks concerning the Universal Coefficient Theorem (UCT) of Rosenberg-Schochet (cf. [73]). A very nice discussion of the UCT can be found in [64]. Indeed, in this manuscript Phillips not only describes what it means for a C*-algebra to satisfy the UCT but he also proves the following two results which we will need.

PROPOSITION 6.2.1 (Two out of Three Principle). *If $0 \to I \to E \to B \to 0$ is a semi-split extension (i.e. there exists a contractive c.p. splitting $B \to E$) and any two of I, E and B satisfy the UCT then so does the third.*

PROPOSITION 6.2.2. *Let $\phi_n : A_n \to A_{n+1}$ be injective $*$-homomorphisms and let A denote the inductive limit of this sequence. Assume further that for each n, there exists a contractive c.p. map $\psi_n : A_{n+1} \to A_n$ such that $\psi_n \circ \phi_n = id_{A_n}$ for all n. If each A_n satisfies the UCT then so does A.*

REMARK 6.2.3. Note that the inductive limit construction used in Theorem 5.3.1 does have the one-sided c.p. inverses required in Proposition 6.2.2 and hence the Popa algebra constructed in that theorem will satisfy the UCT whenever the original residually finite dimensional algebra does. To see that such c.p. inverses exist, we let $\pi : E \to M_k(\mathbb{C})$ be a $*$-homomorphism and $\rho : E \to E \otimes M_n(\mathbb{C})$ be as in the basic construction. The desired map $E \otimes M_n(\mathbb{C}) \to E \cong E \otimes e_{1,1}$ is just given by compressing to the (1,1) corner.

We now answer Lin's question negatively.

THEOREM 6.2.4. *There exists an exact, Popa algebra, A, with real rank zero, stable rank one, UCT, Blackadar's fundamental comparison property (i.e. if $p, q \in A$ are projections such that $\tau(q) < \tau(p)$ for all $\tau \in \mathrm{T}(A)$ then q is equivalent to a subprojection of p), unperforated K-theory, Riesz interpolation property and which is approximately divisible and an increasing union of residually finite dimensional subalgebras such that $\mathrm{T}(A) \neq \mathrm{AT}(A)$.*

PROOF. We first claim that it suffices to find a C*-algebra C which is residually finite dimensional, exact, real rank zero, satisfies the UCT and such that $\mathrm{T}(C) \neq \mathrm{AT}(C)$. Indeed, if we can find such a C then by applying Theorem 5.3.1 to the class of all exact, real rank zero C*-algebras which satisfy the UCT we can find an exact Popa algebra with real rank zero, UCT and such that $\mathrm{T}(A) \neq \mathrm{AT}(A)$. (See Remark 6.2.3 for the UCT assertion and [**16**] for a proof that matrices over a real rank zero algebra also have real rank zero.) Then replacing A with $A \otimes \mathcal{U}$, where $\mathcal{U}$ is some UHF algebra, will be the desired example since this operation preserves Popa's property, exactness, real rank zero, UCT and picks up stable rank one, Riesz interpolation (cf. [**12**, Corollary 3.15]), Blackadar's fundamental comparison property and hence unperforated K-theory (cf. [**70**], [**71**]). Moreover, it is clear that this example will be an inductive limit of residually finite dimensional subalgebras and be approximately divisible in the sense of [**12**].

The construction of the desired residually finite dimensional C*-algebra requires another one of Kirchberg's characterizations of exactness [**46**, Theorem 1.3]: A separable C*-algebra A is exact if and only if there exists a subalgebra B of the CAR algebra, M_{2^∞}, and an AF ideal $J \subset B$ such that $A \cong B/J$. We remark if A is exact and $0 \to J \to B \to A \to 0$ is the short exact sequence given by Kirchberg's theorem, then this sequence is automatically semi-split (i.e. there exists a c.p. splitting $A \to B$) since B is exact and J is nuclear (cf. the bottom of page 41 in [**46**]).

Since $C_r^*(\mathbb{F}_2)$ has a unique trace, it follows from [**71**, Theorem 7.2] that $C_r^*(\mathbb{F}_2) \otimes M_{2^\infty}$ is exact and has real rank zero. By Kirchberg's characterization, we can find an exact, quasidiagonal C*-algebra B with an AF ideal $J \subset B$ such that $B/J \cong C_r^*(\mathbb{F}_2) \otimes M_{2^\infty}$. Moreover, the short exact sequence $0 \to J \to B \to C_r^*(\mathbb{F}_2) \otimes M_{2^\infty} \to 0$ is semi-split.

Since J is AF, it follows from [**16**, Theorem 3.14 and Corollary 3.16] that B also has real rank zero. Note that B has a trace which gives $L(\mathbb{F}_2)\overline{\otimes}R$ in its GNS representation and, by Corollary 4.3.4, this trace is not amenable. Hence $\mathrm{T}(B) \neq \mathrm{AT}(B)$. Finally, since $C_r^*(\mathbb{F}_2)$ satisfies the UCT, it follows from the 'two out of three principle' (cf. Proposition 6.2.1) that B also satisfies the UCT. In other words, B is almost the desired algebra; we only have to replace B with something residually finite dimensional (B is only quasidiagonal).

From the proof of (2) $\Longrightarrow$ (1) in Lemma 6.1.20 we can use B to construct a residually finite dimensional C*-algebra C such that C is exact, real rank zero, satisfies the UCT and such that $\mathrm{T}(C) \neq \mathrm{AT}(C)$. Indeed, that proof produces a short exact sequence

$$0 \to J \to C \to B \to 0,$$

where C is residually finite dimensional, J is an AF algebra (hence C has real rank zero by [**16**, Theorem 3.14 and Corollary 3.16]), the sequence is semi-split (hence

C is exact and satisfies the UCT) and, finally, C will have a non-amenable trace which is gotten by composing the quotient map $C \to B$ with the non-amenable trace on B (since C is locally reflexive and this trace has a non-hyperfinite GNS representation). □

COROLLARY 6.2.5. *There exists an exact, Popa algebra, A, with real rank zero, stable rank one, UCT, unperforated K-theory, Riesz interpolation and Blackadar's fundamental comparison property which is approximately divisible and an increasing union of residually finite dimensional subalgebras but such that A is not tracially AF.*

PROOF. Since $T(A) = UAT(A)_{LFD}$ for every tracially AF algebra (see Proposition 4.5.5), the example in the previous theorem can't be tracially AF. □

It may seem unusual to mention in the previous two results that A is an inductive limit of residually finite dimensional subalgebras. Our main reason for pointing out this fact is that some of Lin's recent structural work on the class of tracially AF C*-algebras relies heavily on a theorem of Blackadar and Kirchberg stating that every simple, nuclear, quasidiagonal C*-algebra is an inductive limit of such subalgebras. In fact, for some of Lin's structural work, this is the only place that nuclearity is used (i.e. his results hold more generally if one replaces the assumption of nuclearity by the assumption of an inductive limit decomposition by residually finite dimensional subalgebras).

REMARK 6.2.6. In [6] Bédos asked whether or not every separable, unital hypertracial C*-algebra is nuclear (see [6, Section 3] - in the language of the present paper, a C*-algebra is hypertracial if every quotient has at least one amenable trace). It is easy to see that every simple, unital, quasidiagonal C*-algebra is hypertracial and hence Dadarlat's examples of non-nuclear Popa algebras provide counterexamples to this question [26]. Theorem 6.2.4 above provides further examples. Indeed, for every non-hyperfinite II_1-factor M which contains a weakly dense exact C*-subalgebra the proof of Theorem 6.2.4 shows that we can construct an exact Popa algebra with stable rank one, Blackadar's comparison property (hence unperforated K-theory), Riesz property, approximate divisibility and which is an increasing union of residually finite dimensional subalgebras but which is not nuclear since it will have $M \bar{\otimes} R$ as the weak closure of some GNS representation.

We now turn to Popa's question. Note that we can't use the examples from Corollary 6.2.5 to produce counterexamples to Popa's question since the GNS representation of any quasidiagonal, locally reflexive (e.g. exact) C*-algebra with unique trace will yield the hyperfinite II_1-factor (which was the obstruction used in the proof of Theorem 6.2.4). Hence to get a Popa algebra with unique trace which is not tracially AF we are forced to step outside the world of exact C*-algebras.

THEOREM 6.2.7. *There exists a quasidiagonal C*-algebra A with real rank zero, stable rank one, strong Dixmier property (i.e. for every $a \in A$ the norm closed, convex hull of $\{uau^* : u \in \mathcal{U}(A)\}$ intersects $\mathbb{C}1_A$) - hence A is simple, has unique trace and is a Popa algebra - but such that A is not tracially AF.*

PROOF. Though the construction is somewhat technical (being quite similar to the inductive limit techniques from Section 5.3), the main idea is not too hard to see when presented abstractly. Our goal is to construct an inductive system

$A_0 \to A_1 \to A_2 \to \cdots$, with connecting maps $\sigma_{n,m} : A_m \to A_n$ and $\{a_k^{(m)}\}_{k \in \mathbb{N}}$ denoting dense sequences in the unit balls of the A_m's, where all of the following conditions are satisfied:

(1) For each n there is a finite set of self-adjoint elements $\mathcal{S}_n \subset (A_n)_{sa}$ such that:
   (a) Each element of $\mathcal{S}_n$ has finite spectrum.
   (b) For each $x \in \sigma_{n,0}(\{a_k^{(0)}\}_{k=1}^n) \cup \ldots \cup \sigma_{n,n-1}(\{a_k^{(n-1)}\}_{k=1}^n)$ there exists an element $y \in \mathcal{S}_n$ such that $\|(x + x*) - y\| < 1/n$.

(2) For each n there is a finite set of invertible elements $\mathcal{I}_n \subset A_n$ such that for each $x \in \sigma_{n,0}(\{a_k^{(0)}\}_{k=1}^n) \cup \sigma_{n,1}(\{a_k^{(1)}\}_{k=1}^n) \cup \ldots \cup \sigma_{n,n-1}(\{a_k^{(n-1)}\}_{k=1}^n)$ there exists an element $y \in \mathcal{I}_n$ with $\|x - y\| < 1/n$.

(3) For each n there is a finite set of unitary elements $\mathcal{U}_n \subset A_n$ such that for each $x \in \sigma_{n,0}(\{a_k^{(0)}\}_{k=1}^n) \cup \sigma_{n,1}(\{a_k^{(1)}\}_{k=1}^n) \cup \ldots \cup \sigma_{n,n-1}(\{a_k^{(n-1)}\}_{k=1}^n)$ there exists a complex number $\lambda(x)$ and positive real numbers $\{\theta_1^{(x)}, \ldots, \theta_{k(x)}^{(x)}\}$ and unitaries $\{u_1, \ldots, u_{k(x)}\} \subset \mathcal{U}_n$ such that $\sum \theta_i^{(x)} = 1$ and
$$\|\lambda(x) 1_{A_n} - \sum_{i=1}^{k(x)} \theta_i^{(x)} u_i x u_i^*\| < 1/n.$$

(4) For each n there is a nonzero, finite dimensional subalgebra $B_n \subset A_n$ with unit e_n such that for each $x \in \sigma_{n,0}(\{a_k^{(0)}\}_{k=1}^n) \cup \sigma_{n,1}(\{a_k^{(1)}\}_{k=1}^n) \cup \ldots \cup \sigma_{n,n-1}(\{a_k^{(n-1)}\}_{k=1}^n)$, $[x, e_n] = 0$ and $e_n x \in B_n$.

(5) For each n, the $*$-homomorphism $\sigma_{n+1,n}$ will be faithful when restricted to the finite dimensional subalgebras $\sigma_{n,0}(B_0), \sigma_{n,1}(B_1), \ldots, \sigma_{n,n-1}(B_{n-1})$, (though the connecting maps will not be injective on all of A_n).

If we are successful in constructing such an inductive system and we let A denote the inductive limit then it is clear that condition (1) will imply that A has real rank zero, (2) will imply stable rank one, (3) will force the strong Dixmier property (which will imply simplicity of A) and thus (by simplicity) (4) and (5) will imply that A is a Popa algebra (hence quasidiagonal – which implies the existence of a tracial state – which is necessarily unique by the Dixmier property). In other words, A satisfies all of the hypotheses stated in the theorem.

In order to deduce that A is not tracially AF, it suffices to show that its unique trace γ is *not* uniform locally finite dimensional. To deduce this, we will also arrange that there exists a C*-algebra C with a trace τ, where τ is known to *not* be uniform amenable, and there exists a sequence of $*$-homomorphisms $\rho_n : C \to A$ such that $\|\gamma \circ \rho_n - \tau\|_{C^*} \to 0$. Since τ is not uniform amenable it will follow that γ can't satisfy the approximation property of a uniform amenable trace either (for if it did then this approximation property would apparently pass to τ). In particular, γ can't have the stronger property of being uniform locally finite dimensional and since every trace on a tracially AF algebra has this stronger property it follows that A can't be tracially AF.

We will use a recursive procedure to construct our inductive system which we now describe abstractly (as we again feel that this is easier to digest). The starting point is to consider a sequence of matrix algebras $\{M_{k(n)}(\mathbb{C})\}$ and a (separable, unital) subalgebra $A_0 \subset \Pi M_{k(n)}(\mathbb{C})$. We claim that if $\mathfrak{F} \subset A_0$ is any finite set then we can find a (separable, unital) C*-algebra $A_1 \subset \Pi M_{l(n)}(\mathbb{C})$ (note that the dimensions of the matrix algebras have changed) and a *-homomorphism $\sigma : A_0 \to A_1$ such that the five properties listed above all hold.

When $A_0 \subset \Pi M_{k(n)}(\mathbb{C})$ is given we can define tracial states τ_n on A_0 by first cutting to the nth summand of $\Pi M_{k(n)}(\mathbb{C})$ and then applying the unique tracial state on $M_{k(n)}(\mathbb{C})$. By compactness, $\{\tau_n\}$ has a weak-$*$ convergent subsequence, and hence for any $\epsilon > 0$ we can find a further subsequence $\{\tau_{n(j)}\}$ such that for every $x \in \mathfrak{F}$, $|\tau_{n(i)}(x) - \lim_{j \to \infty} \tau_{n(j)}(x)| < \epsilon$ for all $i \in \mathbb{N}$. Cutting $\Pi_{n \in \mathbb{N}} M_{k(n)}(\mathbb{C})$ down to the subsequence $\{n(j)\}$ we get a *-homomorphism $\eta : A_0 \to \Pi_{j \in \mathbb{N}} M_{k(n(j))}(\mathbb{C})$. Since von Neumann algebras have the usual Dixmier property (cf. [**44**, Theorem 8.3.5]) we can find a finite set of unitaries $\mathcal{U} \subset \Pi_{j \in \mathbb{N}} M_{k(n(j))}(\mathbb{C})$ such that for each $x \in \mathfrak{F}$ some convex combination of unitary conjugates of x by unitaries from $\mathcal{U}$ will be within ϵ of the center of $\Pi_{j \in \mathbb{N}} M_{k(n(j))}(\mathbb{C})$. Since we have arranged that $|\tau_{n(i)}(x) - \lim_{j \to \infty} \tau_{n(j)}(x)| < \epsilon$ for all $i \in \mathbb{N}$ and $x \in \mathfrak{F}$ this means that we can find positive real numbers $\{\theta_1^{(x)}, \ldots, \theta_{k(x)}^{(x)}\}$ and unitaries $\{u_1, \ldots, u_{k(x)}\} \subset \mathcal{U}$ such that $\sum \theta_i^{(x)} = 1$ and

$$\|(\lim_{j \to \infty} \tau_{n(j)}(x))1 - \sum_{i=1}^{k(x)} \theta_i^{(x)} u_i \eta(x) u_i^*\| < \epsilon.$$

Since $\Pi_{j \in \mathbb{N}} M_{k(n(j))}(\mathbb{C})$ has both real rank zero and stable rank one, we can also find a finite set of self-adjoints $\mathcal{S} \subset \Pi_{j \in \mathbb{N}} M_{k(n(j))}(\mathbb{C})$ and a finite set of invertibles $\mathcal{I} \subset \Pi_{j \in \mathbb{N}} M_{k(n(j))}(\mathbb{C})$ which approximate $\eta(\mathfrak{F})$ appropriately. Letting $D_0 = C^*(\eta(A_0), \mathcal{U}, \mathcal{S}, \mathcal{I})$ we get a (separable, unital) algebra with the first three of the desired approximation properties. At this point it would be most natural to just repeat this procedure, but unfortunately this will not necessarily produce a limit algebra which is quasidiagonal (cf. [**17**, Remark 9.4]). Hence we are forced to add one more complication. Namely, consider the natural finite dimensional representation

$$\pi_{n(1)} : D_0 \to M_{k(n(1))}(\mathbb{C}). \text{ Let } A_1 \subset \left(M_{k(n(1))}(\mathbb{C}) \otimes M_t(\mathbb{C})\right) \otimes \Pi_{j \in \mathcal{N}} M_{k(n(j))}(\mathbb{C})$$

be the C*-algebra generated by $M_{k(n(1))}(\mathbb{C}) \otimes M_t(\mathbb{C})$ and D_0 (here $t > 1/\epsilon$ is just some really big integer) and now define a (unital) *-homomorphism $D_0 \to A_1$ exactly as we did in the basic construction employed in the proof of Theorem 5.3.1 (i.e. replace the natural inclusion by one twisted by the finite dimensional representation $\pi_{n(1)}$). Composing $D_0 \to A_1$ with the map η we get a *-homomorphism $\sigma : A_0 \to A_1$ where we have now arranged properties (1) - (4). Note that (at the next step) condition (5) will be automatic since the finite dimensional $B_1 \cong M_{k(n(1))}(\mathbb{C})$ lives in the left tensor factor of $\left(M_{k(n(1))}(\mathbb{C}) \otimes M_t(\mathbb{C})\right) \otimes \Pi_{j \in \mathcal{N}} M_{k(n(j))}(\mathbb{C}) \cong \Pi_{j \in \mathcal{N}} M_{k(n(1))}(\mathbb{C}) \otimes M_t(\mathbb{C}) \otimes M_{k(n(j))}(\mathbb{C})$. Also note that if we choose the integer t large enough and if τ denotes the weak-$*$ limit of $\{\tau_n\}$ (from the beginning of this paragraph) it is clear from the construction that A_1 has a trace γ such that $\|\tau - \gamma \circ \sigma\|_{A_0^*}$ is as small as you like. Finally, we remark that there is also the natural embedding $D_0 \hookrightarrow A_1 \cong M_{k(n(1))}(\mathbb{C}) \otimes M_t(\mathbb{C}) \otimes D_0$ (which should not be forgotten as we will need it later).

The (actual!) construction of our mystery algebra A begins with any residually finite, discrete, non-amenable group Γ. Let $\Gamma \triangleright \Gamma_1 \triangleright \Gamma_2 \triangleright \cdots$ be a sequence of normal subgroups, each of finite index, such that their intersection is the neutral element. Let $\pi_i : C^*(\Gamma) \to B(l^2(\Gamma/\Gamma_i))$ be the representations induced from the left regular representations of the Γ/Γ_i's and let $\pi = \oplus \pi_i$. A key remark is that if we choose any subsequence $\{i_k\}$ then $\mathrm{tr}_{i_k} \circ \pi_{i_k}(x) \to \tau(x)$ for all $x \in C^*(\Gamma)$, where tr_{i_k} is the unique tracial state on $B(l^2(\Gamma/\Gamma_{i_k}))$ and τ is the trace on $C^*(\Gamma)$ whose GNS representation gives the left regular representation.

Letting $A_0 = \pi(C^*(\Gamma))$ and recursively applying the scheme outlined above, we get a sequence $A_0 \to A_1 \to \cdots$ with all five properties listed at the beginning of the proof and hence we get a limit algebra A satisfying all of the hypotheses of the theorem. Finally, letting $C = C^*(\Gamma)$ it is easy to see that there exists a sequence of $*$-homomorphisms $\psi_n : C \to A_n$ and traces γ_n on A_n such that $\tau = \gamma_n \circ \psi_n$ (use the natural embedding $D_0 \hookrightarrow A_1$ from above) and hence (composing with the natural maps $A_n \to A$) we get a sequence of $*$-homomorphisms $\rho_n : C \to A$ such that $\|\tau - \gamma \circ \rho_n\|_{C^*} \to 0$ (this is basically the same as statement (1) in Theorem 5.3.1). Since Γ was assumed non-amenable (hence τ is not a uniform amenable trace – see Proposition 4.1.6) the proof is now complete. $\square$

REMARK 6.2.8. Note that the examples of the previous theorem also answer Popa's weaker question of whether or not a Popa algebra with unique trace must give the hyperfinite II_1-factor in its GNS representation. Indeed, we deduced that the previous examples were not tracially AF from the fact that their unique traces were not uniform amenable. This is equivalent, by Theorem 3.2.2, to asserting that their GNS constructions do not yield hyperfinite factors and hence there exists a Popa algebra A with unique trace τ such that $\pi_\tau(A)'' \not\cong R$.

6.3. Connes' embedding problem

In this section we show that the techniques of this paper yield new characterizations of those II_1-factors which are embeddable into R^ω. Connes' embedding problem asks whether *every* separable II_1-factor embeds into R^ω. Though it appears to be a technical question only of interest to specialists, it is a remarkable fact that this problem is equivalent to numerous other questions ranging from uniqueness of certain tensor product norms, existence of faithful traces, (local) universality of the trace class operators among all noncommutative L^1-spaces, Kirchberg's 'QWEP conjecture', density of finite dimensional factors (in the Effros-Maréchal topology) in the space of all finite factors and a certain statement about the space of moments of noncommutative random variables in an arbitrary finite W*-probability space. (We highly recommend Ozawa's survey article [58] for precise statements of these equivalences – with a number of simplified proofs – as well as a comprehensive bibliography on the subject.) Moreover, thanks to the deep work of Haagerup [35], a positive answer to Connes' problem would nearly complete the relative invariant subspace problem for II_1-factors while a negative answer would imply a negative answer to Voiculescu's 'unification problem' (i.e. the question of whether or not the microstates and non-microstates approaches to free entropy theory agree). Finally we mention that it seems quite likely that geometric group theory – in particular, Gromov's hyperbolic groups – will play an important role in future developments around this problem. In short, this seemingly technical question is quite deep, intimately related to a surprising variety of other problems and, as such, is a very important open question.

II_1-factors which are embeddable into R^ω already admit a number of characterizations (see [45], [38], [58]) but the results of this section show that the difference between embeddability and hyperfiniteness is rather delicate.

THEOREM 6.3.1. *If $M \subset B(L^2(M))$ is a II_1-factor with trace τ_M then the following are equivalent:*

(1) *M is embeddable into R^ω.*
(2) *There exists a weakly dense C^*-subalgebra $A \subset M$ and a sequence of normal, u.c.p. maps $\varphi_n : M \to M_{k(n)}(\mathbb{C})$ such that*
   (a) *$\|\varphi_n(ab) - \varphi_n(a)\varphi_n(b)\|_2 \to 0$ and*
   (b) *$|\tau_{k(n)} \circ \varphi_n(a) - \tau_M(a)| \to 0$ for all $a, b \in A$.*
(3) *There exists a weakly dense C^*-subalgebra $A \subset M$ and finite rank projections $\{P_n\}$ such that*
   (a) *$\frac{\|[P_n, a]\|_{HS}}{\|P_n\|_{HS}} \to 0$ and*
   (b) *$\frac{\langle aP_n, P_n \rangle_{HS}}{\langle P_n, P_n \rangle_{HS}} \to \tau_M(a)$ for all $a \in A$.*
(4) *There exists a weakly dense C^*-subalgebra $A \subset M$ and a state ϕ on $B(L^2(M))$ such that*
   (a) *$A \subset B(L^2(M))_\phi$[8] and*
   (b) *$\phi|_A = \tau_M$.*
(5) *There exists a weakly dense C^*-subalgebra $A \subset M$ and an embedding $\Phi : M \hookrightarrow R^\omega$ such that $\Phi|_A$ is completely positively liftable.*
(6) *There exists a weakly dense C^*-subalgebra $A \subset M$ such that $C^*(A, JAJ) \cong A \otimes A^{op}$.*
(7) *M has a weak expectation relative to a weakly dense C^*-subalgebra. (i.e. There exists a weakly dense C^*-subalgebra $A \subset M$ and a u.c.p. map $\Phi : B(L^2(M)) \to M$ such that $\Phi|_A = id_A$.)*
(8) *There exists a weakly dense operator system $X \subset M$ such that X is injective.*
(9) *For every finite set $\mathfrak{F} \subset M$ and $\varepsilon > 0$ there exists a complete order embedding $\Phi : R \hookrightarrow M$ (i.e. Φ is an operator system isomorphism from R to $\Phi(R)$) such that for each $x \in \mathfrak{F}$ there exists $r \in R$ such that $\|x - \Phi(r)\|_2 < \varepsilon$.*

PROOF. Our study of representation theory will provide the key to proving this result. Indeed, the first thing we will do is identify $C^*(\mathbb{F}_\infty)$ with a weakly dense subalgebra of M (which is possible thanks to Proposition 5.1.1).

(1) $\implies$ (2). If M embeds into R^ω it follows that $\tau_M|_{C^*(\mathbb{F}_\infty)}$ is an amenable trace on $C^*(\mathbb{F}_\infty)$ (this fact was already used in the proof of Proposition 4.1.14). Hence we can find u.c.p. maps (even $*$-homomorphisms if you like) $\varphi_n : C^*(\mathbb{F}_\infty) \to M_{k(n)}(\mathbb{C})$ which are asymptotically multiplicative (in 2-norm) and recover $\tau_M|_{C^*(\mathbb{F}_\infty)}$ as a weak-$*$ limit after composing with $\text{tr}_{k(n)}$. As we have seen before, one completes the proof by first extending the φ_n's to all of M (via Arveson's theorem) and then replacing the (not necessarily normal) extensions with normal u.c.p. maps (which is possible since there is a one-to-one correspondence between c.p. maps $M \to M_k(\mathbb{C})$ and positive linear functionals on $M_k(M)$ and any positive linear functional can be approximated by normal, positive linear functionals).

Note that statement (2) above is exactly the same as saying that M contains a weakly dense subalgebra $A \subset M$ such that $\tau_M|_A$ is an amenable trace on A. With this reformulation it follows easily from Theorems 3.1.6 and 3.1.7 that (2),

---

[8]Contrary to previous notation, $B(L^2(M))_\phi$ does *not* denote the multiplicative domain. We are following classical notation and now using $B(L^2(M))_\phi$ to denote the centralizer of the state ϕ just as in Corollary 5.3.4.

(3), (4) and (5) are equivalent. Since (5) obviously implies (1) we have shown the equivalence of (1) - (5).

We now wish to add (6) and (7) to the list of equivalent statements so first assume that there exists a weakly dense algebra $A \subset M$ such that $\tau_M|_A$ is an amenable trace on A. By the fourth statement in Theorem 3.1.6 it follows that the natural map $A \odot A^{op} \subset M \odot M^{op} \to B(L^2(M))$ is continuous with respect to the minimal tensor product norm on $A \odot A^{op}$. However, a classical result of Murray and von Neumann states that $M \odot M^{op} \to B(L^2(M))$ is injective and hence $C^*(A, JAJ)$ is necessarily isomorphic to $A \otimes A^{op}$. In other words, (1) $\implies$ (6). That (6) $\implies$ (7) is just another application of Lance's trick (cf. the proof of (4) $\implies$ (5) from Theorem 3.1.6) while that fact that (7) implies $\tau_M|_A$ is an amenable trace on A is already contained in the proof of the last assertion of Theorem 3.1.6. Hence we have shown that (1) - (7) are all equivalent.

(8) $\implies$ (1). In [**45**] Kirchberg proved the following result: $M \subset R^\omega$ if and only if there exists a C*-algebra B with Lance's WEP and a u.c.p. map $\Phi : B \to M$ such that the unit ball of B gets mapped onto a weakly dense subset of the unit ball of M (cf. [**45**, Theorem 1.4] and the equivalence of (vi) and (iii) in [**45**, Proposition 1.3]). Since $B(H)$ is injective it has the WEP and hence this implication follows from Kirchberg's result.

(7) $\implies$ (8). If $A \subset M$ is weakly dense and $\Phi : B(L^2(M)) \to M$ is a u.c.p. map such that $\Phi(a) = a$ for all $a \in A$ then [**9**, Theorem 2.1] ensures that we can find an *idempotent* u.c.p. map $\Psi : B(L^2(M)) \to M$ such that $\Psi(a) = a$ for all $a \in A$. The desired injective operator system is then $X = \Psi(B(L^2(M)))$ (since $\Psi \circ \Psi = \Psi$).

The proof will be complete once we observe the implications (7) $\implies$ (9) and (9) $\implies$ (1).

(7) $\implies$ (9). Given $\mathfrak{F} \subset M$ and $\varepsilon > 0$, let $p \in M$ be a projection whose trace is very close to 1. By the equivalence of (7) and (1), together with Proposition 5.1.1, we can find a weakly dense copy of $C^*(\mathbb{F}_\infty)$ inside (the II_1-factor) pMp such that there exists a u.c.p. map $\Phi : B(L^2(pMp)) \to pMp$ with the property that $\Phi(x) = x$ for all x in the dense copy of $C^*(\mathbb{F}_\infty) \subset pMp$. Since $C^*(\mathbb{F}_\infty)$ also embeds into R we can use Arveson's extension theorem followed by Φ to construct a u.c.p. map $\Psi : R \to pMp$ with weakly dense range. Looking to the orthogonal side, we can also find a unital, normal, embedding $\rho : R \hookrightarrow (1-p)M(1-p)$. It follows that $\Psi \oplus \rho : R \to pMp \oplus (1-p)M(1-p) \subset M$ is a unital, complete order embedding (since ρ is an injective *-homomorphism) whose image nearly contains $\mathfrak{F}$ in 2-norm.

(9) $\implies$ (1). Assuming (9), we will show that one can construct a u.c.p. map

$$\Psi : \Pi_\mathbb{N} R \to M$$

such that the image of the unit ball of $\Pi_\mathbb{N} R$ is weakly dense in the unit ball of M. From Kirchberg's result quoted above it will follow that M is embeddable into R^ω since $\Pi_\mathbb{N} R$ is injective and hence has Lance's WEP. The construction of Ψ is a fairly standard density/ultrafilter argument which goes as follows. Fix a sequence $a_1, a_2, \ldots$ which is weakly dense in the unit ball of M and for each $n \in \mathbb{N}$ let $\phi_n : R \to M$ be a complete order embedding whose range almost contains the first n elements in the sequence $\{a_i\}$ (to within, say, $\frac{1}{n}$). Now fix a free ultrafilter ω and define a u.c.p. map

$$\Psi : \Pi_\mathbb{N} R \to M$$

by declaring
$$\Psi((r_n)_{n\in\mathbb{N}}) = \lim_{n\to\omega} \phi_n(r_n)$$
and it is easily seen that this takes the unit ball onto a weakly dense subset of the unit ball of M. □

REMARK 6.3.2. In [49] Lance introduced the WEP as a property of C*-algebras. In [9] Blackadar changed the point of view to make this a W*-notion as follows: A von Neumann algebra $M \subset B(H)$ has the WEP relative to a weakly dense C*-subalgebra $A \subset M$ if there exists a u.c.p. map $\Phi : B(H) \to M$ such that $\Phi(a) = a$ for all $a \in A$. Lance had earlier speculated that this 'W*-WEP' would actually imply injectivity (indeed, a von Neumann algebra has the 'C*-WEP' if and only if it is injective) however a (non-factor) counterexample was given in [9]. In [10] and [45, Corollary 3.5] it was shown that there exist non-hyperfinite *factors* with the W*-WEP, but the proofs were non-constructive and concrete examples remained elusive. In a preliminary version of this paper we gave the first concrete examples of non-injective factors (actually McDuff II_1-factors) with the W*-WEP. Shortly after that, in joint work with Dykema [19], it was shown that all the (interpolated) free group factors on a finite number of generators have the W*-WEP. However, the previous theorem shows that most of the standard examples of II_1-factors (e.g. things coming from residually finite groups like free groups, $SL(n,\mathbb{Z})$ or $\mathbb{Z}^2 \rtimes SL(2,\mathbb{Z})$) have the W*-WEP and hence we get examples with property T and even factors with trivial fundamental group [68].

In other words, the W*-WEP is not an exotic property and it is enjoyed by many II_1-factors which are quite far from being hyperfinite. On the other hand, we strongly believe that this notion is still 'too close' to injectivity to expect that *every* II_1-factor has this property. For example, from part (9) of Theorem 6.3.1 we get that a II_1-factor has the W*-WEP if and only if it is quite literally built out the hyperfinite II_1-factor in a way which naturally mixes von Neumann algebraic and operator space notions. Our feeling is that this is not that far (*very* loosely speaking) from being injective – far enough to exhibit vastly different properties like no Cartan subalgebras, property T, trivial fundamental group, etc. – but, still, not that far.

In order to illustrate just how delicate the results above are, we remind the reader of the various characterizations of the hyperfinite II_1-factor (most of which are due to Alain Connes).

THEOREM 6.3.3. *If $M \subset B(L^2(M))$ is a II_1-factor with trace τ_M then the following are equivalent:*

(1) $M \cong R$.
(2) *There exists a weakly dense C*-subalgebra $A \subset M$ and a sequence of normal, u.c.p. maps $\varphi_n : M \to M_{k(n)}(\mathbb{C})$ such that*
   (a) $\|\varphi_n(ab) - \varphi_n(a)\varphi_n(b)\|_2 \to 0$ *for all $a, b \in A$ and*
   (b) $|\tau_{k(n)} \circ \varphi_n(x) - \tau_M(x)| \to 0$ *for all $x \in M$.*
(3) *There exists a weakly dense C*-subalgebra $A \subset M$ and finite rank projections $\{P_n\}$ such that*
   (a) $\frac{\|[P_n,a]\|_{HS}}{\|P_n\|_{HS}} \to 0$ *for every $a \in A$ and*
   (b) $\frac{\langle xP_n,P_n\rangle_{HS}}{\langle P_n,P_n\rangle_{HS}} \to \tau_M(x)$ *for all $x \in M$.*

(4) There exists a weakly dense C*-subalgebra $A \subset M$ and a state ϕ on $B(L^2(M))$ such that
   (a) $A \subset B(L^2(M))_\phi$ and
   (b) $\phi|_M = \tau_M$.
(5) There exists a completely positively liftable embedding $\Phi : M \hookrightarrow R^\omega$.
(6) $C^*(M, JMJ) \cong M \otimes M^{op}$.
(7) There exists a u.c.p. map $\Phi : B(L^2(M)) \to M$ such that $\Phi|_M = id_M$ (i.e. M is injective).
(8) For every finite set $\mathfrak{F} \subset M$ and $\varepsilon > 0$ there exists a *-monomorphism $\Phi : R \hookrightarrow M$ such that for each $x \in \mathfrak{F}$ there exists $r \in R$ such that $\|x - \Phi(r)\|_2 < \varepsilon$.

PROOF. (1) $\Longrightarrow$ (2) is obvious. (2) $\Longrightarrow$ (5) is contained in the argument in the proof of (1) $\Longrightarrow$ (2) from Theorem 3.2.2. Since (5) is equivalent to hyperfiniteness for general finite von Neumann algebras, we see that (1), (2) and (5) are equivalent.

The equivalence of (1), (6) and (7) is due to Connes (cf. [**23**, Theorem 5.1]). This paper also contains his adaptation of Day's trick to deduce (3) from injectivity (7). Note also that the equivalence of (8) and (1) goes all the way back to Murray and von Neumann. Hence we are left to prove (3) $\Longrightarrow$ (4) and (4) $\Longrightarrow$ (1).

(3) $\Longrightarrow$ (4) is well known: simply take a cluster point of the states
$$T \mapsto \frac{\langle TP_n, P_n \rangle_{HS}}{\langle P_n, P_n \rangle_{HS}}$$
on $B(L^2(M))$. Finally, (4) $\Longrightarrow$ (1) also follows from Connes' work since the density of A in M, together with the fact that the hypertrace takes the correct value on all of M, implies that actually M is contained in the centralizer of φ and hence is hyperfinite (cf. proof of [**24**, Theorem 5.1]). $\square$

We close this section with a simple result relating amenable traces on C*-algebras with the *local lifting property* (LLP) and Connes' embedding problem. We have already seen that in many instances every trace on a particular C*-algebra will enjoy some type of amenability (e.g. every trace on a C*-algebra with the WEP is amenable). We now observe that Connes' embedding problem predicts that every trace on a C*-algebra with the LLP is amenable and hence it is possible that a counterexample to Connes' problem could be constructed by finding a non-amenable trace on some C*-algebra with the LLP.

By definition, A has the local lifting property if for every C*-algebra B, ideal $J \triangleleft B$, u.c.p. map $\phi : A \to B/J$ and finite dimensional operator system $X \subset A$ there exists a u.c.p. lifting of $\phi|_X$.

PROPOSITION 6.3.4. *Assume that A has the LLP and $\tau \in T(A)$. Then τ is amenable if and only if there exists a τ''-preserving embedding*
$$\pi_\tau(A)'' \hookrightarrow R^\omega,$$
where τ'' denotes the normal extension of τ to $\pi_\tau(A)''$.

PROOF. This is very similar to the equivalence of statements (1) and (2) from Theorem 3.1.7 and hence the details are left to the reader. $\square$

For all intents and purposes the following result is due to Kirchberg (cf. [**45**]) however the proof is very simple so we include it.

PROPOSITION 6.3.5. *The following statements are equivalent.*
(1) *Every (separable)* II_1-*factor embeds into* R^ω.
(2) $T(A) = AT(A)$ *for every A with the LLP.*
(3) $T(C^*(\mathbb{F}_\infty)) = AT(C^*(\mathbb{F}_\infty))$.

PROOF. (1) $\Longrightarrow$ (2). Since AT(A) is always a face in T(A) it follows that every trace on A is amenable if and only if every factorial trace on A is amenable. With this observation and Proposition 6.3.4 in hand, it is easily seen that an affirmative answer to Connes' problem would imply (2).

(2) $\Longrightarrow$ (3). In [**46**, Lemma 3.3] Kirchberg showed that $C^*(\mathbb{F}_\infty)$ has the LLP.

(3) $\Longrightarrow$ (1). We have already observed that every II_1-factor arises as the GNS representation of $C^*(\mathbb{F}_\infty)$ with respect to some trace and hence this part follows from Theorem 3.1.7. □

6.4. Amenable traces and numerical analysis

In this section we observe that the results and techniques of these notes are relevant to some natural and important questions which lie in the intersection of operator theory and numerical analysis. In order to stay reasonably self-contained we will recall the definitions and (statements of) results that we need, however we highly recommend looking at the recent book of Hagen, Roch and Silbermann [**40**] (see also Böttcher's survey article [**15**]) for more on this subject.

Let $T \in B(H)$ be given. There are two important questions that one may hope to (approximately) solve via a combination of operator theoretic and numerical analytic techniques. Namely, (1) given a vector $v \in H$ find a vector $w \in H$ such that $Tw = v$ and (2) compute the spectrum, $\sigma(T)$, of T. (In fact, there are several other questions one may ask and we will see another concerning spectral distributions of self-adjoint operators later in this section. However, for now, we will stick to these two basic problems.) If H happens to be a finite dimensional Hilbert space then, in many instances, there are efficient algorithms for attacking these problems numerically. For infinite dimensional H the 'finite section method' provides a natural *strategy* for reducing these questions to the finite dimensional case where one can then apply the established numerical algorithms. Of course, the real question – at least from a theoretical point of view – is whether or not this strategy will work.

To make this more precise, we need to introduce some terminology. A *filtration* of H is an increasing sequence of finite rank projections $P_1 \leq P_2 \leq \cdots$ such that $\|P_n(x) - x\| \to 0$, as $n \to \infty$, for every vector $x \in H$. The *finite section method* is to take an operator $T \in B(H)$, cut it down to a filtration (i.e. consider the sequence $P_n T P_n$ – now regarded as acting on finite dimensional Hilbert spaces) and hope that by solving questions (1) and (2) – presumably numerically – for each $P_n T P_n$ we can pass to a limit to recover the solutions for T. While this idea could not be more natural, great care must be taken in choosing the 'right' filtration for a particular operator T as it can easily happen that $P_n T P_n$ do not provide good approximations to T. (Think of cutting down the bilateral shift on $l^2(\mathbb{Z})$ to a filtration defined by the canonical basis. The spectrum of the bilateral shift is $\mathbb{T}$ while the cut-downs will all be nilpotent and hence have spectrum 0.)

In relation to solving general vector equations, the following definition naturally arises: we say that the finite section method (w.r.t. a given filtration) is *stable* (for

T) if there exists a number n_0 such that for all $n \geq n_0$, $P_n T P_n$ are invertible (as operators from $P_n H \to P_n H$) and

$$\sup_{n \geq n_0} \|(P_n T P_n)^{-1}\| < \infty.$$

The following result summarizes the usefulness of this notion (cf. [40, Theorem 1.4], [15, Proposition 1.2]).

THEOREM 6.4.1. *Let $T \in B(H)$ and a filtration $P_1 \leq P_2 \leq \cdots$ be given. If the finite section method is stable then:*

(1) *T is invertible.*
(2) *Given a vector $v \in H$, let $v_n = P_n v$. Then, for sufficiently large n, there exists a (necessarily unique) vector $w_n \in P_n H$ such that $(P_n T P_n) w_n = v_n$ and, moreover, $\{w_n\}$ is a norm convergent sequence (in H) with limit w, where w is the (necessarily unique) solution to the equation $Tw = v$.*

In other words, stability is the notion which ensures that there is hope of solving (infinite dimensional) equations numerically.

To handle the spectral approximation problem we will need more terminology.

DEFINITION 6.4.2. (1) If X_n is a sequence of subsets of the complex plane then we let $\limsup_{n \to \infty} X_n$ (resp. $\liminf_{n \to \infty} X_n$) denote the set of all cluster points (resp. limits) of sequences $\{x_n\}$ with $x_n \in X_n$. Another way of saying this is that $\liminf_{n \to \infty} X_n$ is the set of limits along all possible sequences of points in $\{X_n\}$ while $\limsup_{n \to \infty} X_n$ is the set of limits along all possible subsequences of points. Evidently we always have $\liminf_{n \to \infty} X_n \subset \limsup_{n \to \infty} X_n$. (An example: Let $X_n = \{(-1)^n + 1/n, 0\}$. Then $\liminf_{n \to \infty} X_n = \{0\}$ while $\limsup_{n \to \infty} X_n = \{-1, 0, 1\}$.)

(2) Given $S \in B(H)$ and $\varepsilon > 0$, the *ε-pseudospectrum of S*, denoted $\sigma^{(\varepsilon)}(S)$, is the union of the usual spectrum, $\sigma(S)$, together with the set of points $\lambda \in \mathbb{C} \setminus \sigma(S)$ such that $\|(\lambda - S)^{-1}\| \geq 1/\varepsilon$. Note that if $\varepsilon_1 \leq \varepsilon_2$ then $\sigma^{(\varepsilon_1)}(S) \subset \sigma^{(\varepsilon_2)}(S)$ and, also,

$$\sigma(S) = \bigcap_{\varepsilon > 0} \sigma^{(\varepsilon)}(S).$$

The first definition above is simply a precise formulation of the notion of convergence of spectra which will appear in our results. The notion of pseudospectrum is relevant to our discussion because it turns out that these sets behave better than actual spectra when passing to limits. More precisely, we will need the following result.

THEOREM 6.4.3. [40, Theorem 3.31] *Let $(T_n) \in \Pi M_{k(n)}(\mathbb{C})$ and $\varepsilon > 0$ be given. Then*

$$\sigma^{(\varepsilon)}((T_n) + \oplus M_{k(n)}(\mathbb{C})) = \limsup_{n \to \infty} \sigma^{(\varepsilon)}(T_n),$$

where $(T_n) + \oplus M_{k(n)}(\mathbb{C})$ is the image of (T_n) in the quotient algebra $\Pi M_{k(n)}(\mathbb{C}) / \oplus M_{k(n)}(\mathbb{C})$.

For actual spectra one always has the inclusion
$$\sigma((T_n) + \oplus M_{k(n)}(\mathbb{C})) \supset \limsup_{n \to \infty} \sigma(T_n),$$
but the inclusion can be proper. However, it is an interesting fact (cf. [**40**, Corollary 3.8]) that if each matrix T_n is normal then we get the equation $\sigma((T_n) + \oplus M_{k(n)}(\mathbb{C})) = \limsup_{n \to \infty} \sigma(T_n)$.

For an arbitrary element T in a C*-algebra we will let $C^*(T)$ denote the unital C*-algebra generated by T. Also, if $\{P_n\}$ is a filtration of H and $T \in B(H)$ then φ_n will denote the state on $C^*(T)$ given by $\varphi_n(X) = \text{tr}_{rank(P_n)}(P_n X P_n)$ for all $X \in C^*(T)$.

We are finally in a position to state the main theorem of this section.

THEOREM 6.4.4. *Let $T \in B(H)$ be a quasidiagonal operator and assume that the C*-algebra generated by T is exact. Then there exists a filtration $P_1 \leq P_2 \leq \cdots$ such that:*

(1) *The finite section method is stable if and only if T is invertible.*
(2) $\sigma(T) = \bigcap_{\varepsilon > 0} \left(\limsup_{n \to \infty} \sigma^{(\varepsilon)}(P_n T P_n) \right).$
(3) *For every $\tau \in \text{T}(\text{C}^*(\pi(\text{T})))_{\text{QD}}$ there exists a subsequence $\{n(k)\}$ such that $\varphi_{n(k)} \to \tau \circ \pi|_{C^*(T)}$ in the weak-* topology, where $\pi : B(H) \to Q(H)$ is the quotient mapping onto the Calkin algebra.*

PROOF. We first claim that it will suffice to prove the following local statement: Given $\tau \in \text{T}(\text{C}^*(\pi(\text{T})))_{\text{QD}}$, a finite set $\mathfrak{F} \subset C^*(T)$, a finite set $F \subset H$ and $\delta > 0$ there exists a finite rank projection $P \in B(H)$ such that:

(a) $\|[P, X]\| < \delta$ for all $X \in \mathfrak{F}$.
(b) $Px = x$ for all $x \in F$.
(c) $|\text{tr}_{rank(P)}(PXP) - \tau(\pi(X))| < \delta$ for all $X \in \mathfrak{F}$.

Assuming that this local statement holds, an argument similar to the one given in the proof of Proposition 3.3.2 would show that one can construct a filtration $P_1 \leq P_2 \leq \cdots$ such that $\|[P_n, T]\| \to 0$ and for every $\tau \in \text{T}(\text{C}^*(\pi(\text{T})))_{\text{QD}}$ there exists a subsequence $\{n(k)\}$ such that $\varphi_{n(k)} \to \tau \circ \pi|_{C^*(T)}$ in the weak-* topology. This evidently yields part (3) of the theorem. However it also gives statements (1) and (2). Indeed, letting $\Phi : C^*(T) \to \Pi P_n B(H) P_n / \oplus P_n B(H) P_n$ be the u.c.p. map which takes X to $P_n X P_n + \oplus P_n B(H) P_n$ we have that in fact Φ is a faithful *-homomorphism. The faithfulness follows from the fact that the P_n's are a filtration and hence $\|X\| = \lim \|P_n X P_n\|$ for every $X \in B(H)$. That Φ is multiplicative follows from the quasidiagonality assumption since
$$\|P_n XY P_n - P_n X P_n Y P_n\| = \|P_n X (Y P_n - P_n Y) P_n\| \leq \|X\| \|[P_n, Y]\| \to 0.$$

From the theorem quoted above it follows that for every $\epsilon > 0$,
$$\sigma^{(\varepsilon)}(\Phi(T)) = \limsup_{n \to \infty} \sigma^{(\varepsilon)}(P_n T P_n).$$

Since $\sigma(S) = \cap \sigma^{(\varepsilon)}(S)$ holds in general and since Φ is a faithful homomorphism we conclude that
$$\sigma(T) = \sigma(\Phi(T)) = \cap \sigma^{(\varepsilon)}(\Phi(T)) = \cap \limsup_{n \to \infty} \sigma^{(\varepsilon)}(P_n T P_n).$$

To see that part (1) also follows we observe that the finite section method is stable if and only if $(P_n T P_n) + \oplus P_n B(H) P_n$ is an invertible element of $\Pi P_n B(H) P_n / \oplus$

$P_n B(H) P_n$ (this is a straightforward exercise – or see [**40**, Theorem 1.15]). Hence stability is equivalent to the invertibility of $\Phi(T)$ which, by preservation of spectra, is equivalent to the invertibility of T.

The proof of the required local statement is also similar to the proof of Proposition 3.3.2 however there are a few additional wrinkles to iron out. First of all, since $C^*(T)$ is exact (hence locally reflexive), we can apply the Effros-Haagerup lifting theorem (cf. [**81**]) to find a u.c.p. splitting $\Phi : C^*(\pi(T)) \to B(H)$. It follows that $C^*(\pi(T))$ is also an exact, quasidiagonal C*-algebra. Indeed, exactness always passes to quotients but we also get quasidiagonality from Voiculescu's abstract characterization since we can compose the splitting Φ with compression by an appropriately chosen sequence of finite rank projections (coming from the quasidiagonality of the operator T) to produce some u.c.p. maps to matrix algebras which are asymptotically multiplicative in norm and asymptotically isometric.

Since T is a quasidiagonal operator, if we are given finite sets $\mathfrak{F} \subset C^*(T)$ and $F \subset H$ we can find a finite rank projection Q such that $Qx = x$ for all $x \in F$, $\|[Q, X]\|$ is as small as we like and (hence) $\|X - (QXQ + Q^\perp X Q^\perp)\|$ is also as small as desired for every operator $X \in \mathfrak{F}$. Note that the u.c.p. map $\Phi_{Q^\perp} : C^*(\pi(T)) \to B(Q^\perp H)$, $\Phi_{Q^\perp}(Y) = Q^\perp \Phi(Y) Q^\perp$ is still a u.c.p. splitting (i.e. a faithful *-homomorphism modulo the compacts) and, moreover, is nearly multiplicative on $\pi(\mathfrak{F})$. Letting $\rho : C^*(\pi(T)) \to B(K)$ be any faithful representation which contains no non-trivial compact operators, we can, thanks to Proposition 3.3.2, find a filtration $Q_1 \leq Q_2 \leq \cdots$ which asymptotically commutes (in norm) with $\rho(C^*(\pi(T)))$ and, furthermore, which recaptures any fixed quasidiagonal trace $\tau \in \mathrm{T}(C^*(\pi(T)))_{\mathrm{QD}}$. From Voiculescu's Theorem (version 2.3) we can find a unitary $U : K \to Q^\perp H$ such that $\|Q^\perp X Q^\perp - U\rho(\pi(X))U^*\|$ is very small for all $X \in \mathfrak{F}$. Defining $P_n = Q \oplus UQ_n U^*$ we get a filtration of H where each projection almost commutes with $\mathfrak{F}$ (they are *not* asymptotically commuting, but there is a small, uniform upper bound on the norms of commutators) and cutting by these projections will almost recover τ (since Q is fixed and finite dimensional, its contribution becomes negligible as $n \to \infty$). $\square$

REMARK 6.4.5. In most cases, part (3) above is a much stronger statement than part (2). Indeed, in [**5**] Arveson observes how statement (3), which he regards as an analogue of Szegő's Limit Theorem, can be used to recover the essential spectrum of a self-adjoint operator.

Note that we only used exactness of C*(T) to deduce the existence of a u.c.p. splitting C*$(\pi(T)) \to B(H)$ (which, in turn, was only needed in part (3) above). Hence the result above holds under this weaker assumption. However, many standard examples in operator theory are easily seen to generate exact C*-algebras. For example, essentially normal operators (hence Toeplitz operators with continuous symbol) and all weighted shift operators have this property (since essentially normal operators generate nuclear C*-algebras while every weighted shift is an element of the nuclear C*-algebra $l^\infty(\mathbb{N}) \rtimes \mathbb{N}$ and hence generates something exact). Many of the concrete operators considered in [**40**] can be realized as weighted shifts (or, more generally, as elements in one of the algebras $l^\infty(\mathbb{N}) \rtimes \mathbb{N}$ or $l^\infty(\mathbb{Z}) \rtimes \mathbb{Z}$) and hence generate exact C*-algebras. Of course, quasidiagonality of concrete examples is a much harder question. However for the classes of essentially normal and (finite direct sums of) weighted shift operators there are nice characterizations of quasidiagonality as well (cf. [**27**, Corollary IX.7.4, Section IX.8], [**74**], [**55**]).

Moreover, we should also point out that if a weighted shift satisfies the appropriate hypotheses (cf. [**74**]) then it is possible to actually construct (as opposed to proving the existence of) an asymptotically commuting filtration via Berg's technique (this would be enough for statements (1) and (2) above, but not necessarily enough for conclusion (3)).

Specializing to the case of a self-adjoint operator Theorem 6.4.4 takes an especially nice form. Note that if $T \in B(H)$ is self-adjoint and $P \in B(H)$ is a projection then PTP is still a self-adjoint operator (in particular, it is normal – see the remark after Theorem 6.4.3). Hence if $\{P_n\}$ is a filtration and the rank of P_n is some integer d_n then $P_n T P_n$ will have d_n (not necessarily distinct) eigenvalues which we will denote by $\lambda_1^{(n)}, \ldots, \lambda_{d_n}^{(n)}$. To complete the spectral approximation picture we will need one more of Arveson's notions: A real number λ will be called *essential* (for T with respect to $\{P_n\}$) if for every open interval U containing λ we have

$$\lim_{n \to \infty} N_n(U) = \infty,$$

where $N_n(U)$ is the number of eigenvalues of $P_n T P_n$ (counted with multiplicities) in the interval U. We then let $\liminf_{ess} \sigma(P_n T P_n)$ denote the set of essential points. Note that $\liminf_{ess} \sigma(P_n T P_n) \subset \liminf \sigma(P_n T P_n)$.

COROLLARY 6.4.6. *Let $T \in B(H)$ be a self-adjoint operator. Then there exists a filtration $P_1 \leq P_2 \leq P_3 \cdots$ such that:*
(1) *The finite section method is stable if and only if T is invertible.*
(2) $\sigma(T) = \limsup_{n \to \infty} \sigma(P_n T P_n) = \liminf_{n \to \infty} \sigma(P_n T P_n)$.
(3) $\sigma_{ess}(T) = \liminf_{ess} \sigma(P_n T P_n)$.
(4) *If μ is a regular, Borel, probability measure on $\sigma(T)$ then there exists a subsequence $\{n(k)\}$ such that for every $f \in C(\sigma(T))$ we have*

$$\lim_{k \to \infty} \frac{f(\lambda_1^{(n(k))}) + f(\lambda_2^{(n(k))}) + \cdots + f(\lambda_{d_{n(k)}}^{(n(k))})}{d_{n(k)}} = \int_{\sigma(T)} f(x) d\mu(x)$$

if and only if μ is supported on the essential spectrum of T.

PROOF. Let $P_1 \leq P_2 \leq \cdots$ be the filtration given by Theorem 6.4.4. According to [**40**, Theorem 7.2] one always has the inclusion $\sigma(T) \subset \liminf \sigma(P_n T P_n)$ – so long as T is self-adjoint and $\{P_n\}$ is a filtration – and hence we get

$$\sigma(T) \subset \liminf \sigma(P_n T P_n) \subset \limsup \sigma(P_n T P_n) = \sigma(T),$$

where we have used the fact that $P_n T P_n$ are normal and hence the ε-psuedospectra in Theorem 6.4.4 are unnecessary (see the remark after Theorem 6.4.3).

Conclusion (3) is essentially due to Arveson. The set-up here is not exactly the same as that in [**40**, Theorem 7.10] but the difference is minor and the details will be left to the reader.

For the final statement above first note that there is a one-to-one correspondence between regular, Borel, probability measures on the essential spectrum of T and quasidiagonal traces on $C^*(\pi(T))$ (since every state on an abelian C*-algebra is, in fact, a uniform locally finite dimensional trace). The conclusion now follows from the corresponding statement in Theorem 6.4.4. □

REMARK 6.4.7. In [**5**] Arveson proved the essential parts of the corollary above under the following additional hypotheses: T should be a self-adjoint element of a simple, unital C*-algebra A which has a unique tracial state and which admits an 'A-filtration' (see [**5**, Theorems 3.8 and 4.5]). Of course, at the time Arveson was introducing these notions and the point of his results was to show that there was an abstract set of hypotheses which would ensure convergence results (and hence give hope of attacking infinite dimensional problems numerically). Moreover, he showed that a number of interesting examples where covered by his results. In the remarks after [**5**, Proposition 4.4] Arveson notes that it is natural to ask which C*-algebras admit an 'A-filtration' (hence fall under his results) and points out that things like Cuntz algebras do not (the existence of an 'A-filtration' implies the existence of a tracial state – which no purely infinite C*-algebra will have). However, from the theoretical point of view, it does not follow that self-adjoint operators in Cuntz algebra are beyond the reach of these techniques. Indeed, the corollary above holds for a completely arbitrary self-adjoint operator and hence there is *always* hope of attacking infinite dimensional self-adjoint operators numerically.

6.5. Amenable traces and obstructions in K-homology

In this section we will show how the theory of amenable traces can be used to solve some K-homological questions and a natural question regarding the passage of quasidiagonality to the Calkin algebra. The main results of this section are taken from [**2**], [**24**], [**79**] and [**80**]. However, the invariant mean point of view will turn these "technical" results into very natural and intuitively clear theorems.

The three questions which are easily solved using amenable traces are the following:
  (1) (cf. [**24**]) If Γ is a discrete group can one always construct a finitely summable, unbounded Fredholm module over $C_r^*(\Gamma)$?
  (2) (cf. [**2**], [**79**], [**80**]) Is the BDF semigroup Ext($\cdot$) always a group?[9]
  (3) (cf. [**79**], [**80**]) If $A \subset B(H)$ is a quasidiagonal set of operators is it true that the image of A in the Calkin algebra is quasidiagonal?

The answer to all the questions above is 'No'. It is intriguing, however, that there is a single reason why all three questions are false: *Not all traces are amenable.*

This obstruction is explicitly pointed out in [**24**] and implicitly used in [**2**], [**79**] and [**80**]. Indeed we can reformulate [**24**, Lemma 9] as follows.

LEMMA 6.5.1. *(See [**24**, Lemma 9, Remark 10(b)]) Let A be a C*-algebra and assume that A admits a finitely summable, unbounded Fredholm module. Then* $\mathrm{AT}(A) \neq \emptyset$.

Since the *reduced* group C*-algebra of any discrete, non-amenable group has no amenable traces (cf. Proposition 4.1.1) we immediately deduce the following result.

COROLLARY 6.5.2. *(cf. [**24**, Theorem 19]) Let Γ be any discrete, non-amenable group. Then $C_r^*(\Gamma)$ has no unbounded, finitely summable Fredholm modules.*

REMARK 6.5.3. In [**76**] Voiculescu defined a notion of 'subexponential', unbounded Fredholm module and generalized Connes' result to this setting (i.e. the

---

[9]See [**27**] for a nice introduction to Ext and BDF theory. In particular we refer to the same text for the definitions and basic facts regarding the extension semigroup. All we will need in this paper is the following result of Arveson [**3**]: For a given C*-algebra A, Ext(A) is *not* a group if and only if there exists a *-homomorphism $\pi : A \to Q(H)$ to the Calkin algebra with the property that there is *no* u.c.p. lifting into $B(H)$.

reduced group C*-algebra of a non-amenable, discrete group never admits one of these either – see [**76**, Proposition 4.10]). Amenable traces explain this result in exactly the same way: If a C*-algebra A admits a subexponential, unbounded Fredholm module then $\mathrm{AT}(A) \neq \emptyset$ (cf. [**76**, Proposition 4.6]) and hence $C_r^*(\Gamma)$ has no such modules whenever Γ is not amenable.

The Ext and quasidiagonality questions above are easily handled simultaneously. We believe that the ideas involved become most transparent when treated in generality and so this is what we will do. The main idea is to construct a C*-algebra which has a tracial state which is not amenable. The easiest example of such an algebra is the *reduced* group C*-algebra of any discrete, non-amenable group. Unfortunately these algebras don't quite do the trick, but certain extensions of them will.[10]

We begin with some general remarks which underlie the main ideas. Let $M_{k(n)}(\mathbb{C})$ be a sequence of matrix algebras and

$$\Pi M_{k(n)} = \{(x_n) : \sup_n \|x_n\| < \infty\}$$

denote the l^∞-product. The c_0-product,

$$\oplus M_{k(n)} = \{(x_n) : \|x_n\| \to 0, n \to \infty\},$$

sits as an ideal in $\Pi M_{k(n)}$ and we will identify the quotient C*-algebra,

$$\Pi M_{k(n)} / \oplus M_{k(n)},$$

with a subalgebra of the Calkin algebra (associated to the Hilbert space $H = \oplus \mathbb{C}^{k(n)}$). Note that if $\omega \in \beta\mathbb{N}\backslash\mathbb{N}$ is a free ultrafilter then we can define a trace τ_ω on $\Pi M_{k(n)} / \oplus M_{k(n)}$ by

$$\tau_\omega((x_n) + \oplus M_{k(n)}) = \lim_{n \to \omega} \mathrm{tr}_{k(n)}(x_n).$$

The following lemma explains why Ext need not be a group and why quasidiagonality need not pass to the Calkin algebra.

LEMMA 6.5.4. *Let* $A \subset \Pi M_{k(n)} / \oplus M_{k(n)}$ *be given and assume that* $\tau_\omega|_A$ *is not an amenable trace on* A. *Then* $\mathrm{Ext}(A)$ *is not a group. If* A *has no amenable traces whatsoever then* A *is not quasidiagonal.*

PROOF. Assume that $\mathrm{Ext}(A)$ is a group. Then there exists a u.c.p. map $\Phi : A \to B(H)$ which is a splitting for the inclusion $A \hookrightarrow \Pi M_{k(n)} / \oplus M_{k(n)} \subset Q(H)$ (cf. [**3**]). Recall that $H = \oplus \mathbb{C}^{k(n)}$ and we have a natural inclusion $\Pi M_{k(n)} \subset B(H)$ and a natural conditional expectation $E : B(H) \to \Pi M_{k(n)}$ given by

$$E(T) = (P_n T P_n)_{n \in \mathbb{N}}$$

where P_n denotes the unit of $M_{k(n)}$. This conditional expectation evidently has the property that it maps compact operators into compact operators (i.e. into $\oplus M_{k(n)}$) and this implies that $E \circ \Phi : A \to \Pi M_{k(n)}$ is also a splitting. Indeed, since each $a \in A$ has a lift, say $\tilde{a}$, in $\Pi M_{k(n)}$ we have that $\Phi(a) - \tilde{a}$ is compact and hence

$$E(\Phi(a) - \tilde{a}) = E \circ \Phi(a) - \tilde{a} \in \oplus M_{k(n)}.$$

This implies that $E \circ \Phi(a)$ is also a lift of a for each $a \in A$.

---

[10]Actually, in [**37**] Haagerup and Thorbjørnsen have recently shown that $C_r^*(\mathbb{F}_2)$ will do the trick! They have shown the remarkable fact that $C_r^*(\mathbb{F}_2) \subset \Pi M_{k(n)} / \oplus M_{k(n)}$ and hence Lemma 6.5.4 applies.

To ease notation, we will now forget about E and just let $\Phi : A \to \Pi M_{k(n)}$ be a u.c.p. map which lifts the natural inclusion $A \subset Q(H)$. For each $n \in \mathbb{N}$ we now define a u.c.p. map $\phi_n : A \to M_{k(n)}$ by

$$\phi_n(a) = P_n \Phi(a) P_n.$$

Since Φ is a splitting we have that $\Phi(ab) - \Phi(a)\Phi(b) \in \oplus M_{k(n)}$ and this implies that the maps ϕ_n are asymptotically multiplicative in norm. Also, it is not hard to see that $\tau_\omega|_A$ is in the weak-* closure of the states $\{\mathrm{tr}_{k(n)} \circ \phi_n\} \subset S(A)$ and this implies that $\tau_\omega|_A$ is a quasidiagonal trace on A (hence amenable) which contradicts our assumption.

The second statement above is trivial since every (unital) quasidiagonal C*-algebra has at least one quasidiagonal tracial state. □

Thus our strategy is clear – construct some algebra $A \subset \Pi M_{k(n)}/\oplus M_{k(n)}$ such that $\tau_\omega|_A$ is not an amenable trace on A or, better yet, so that A has no amenable traces at all. There are two ways of using discrete groups to realize this strategy. One starts from a discrete, non-amenable, residually finite group and the other starts from a discrete, property T group with infinitely many non-equivalent, finite dimensional, irreducible, unitary representations. We treat the former case first.

So let Γ be a non-amenable, residually finite group (e.g. a free group) and $\Gamma \triangleright \Gamma_1 \triangleright \Gamma_2 \triangleright \cdots$ be a descending sequence of normal subgroups of finite index such that the intersection of the Γ_n's is the neutral element. Let $\pi_n : C^*(\Gamma) \to B(l^2(\Gamma/\Gamma_n)) \cong M_{k(n)}$ be the representations induced by the left regular representations of the Γ/Γ_n's. Define

$$\pi = \oplus \pi_n : C^*(\Gamma) \to \Pi M_{k(n)}.$$

Now consider the algebra

$$B = \sigma \circ \pi(C^*(\Gamma)) \subset \Pi M_{k(n)}/\oplus M_{k(n)},$$

where $\sigma : \Pi M_{k(n)} \to \Pi M_{k(n)}/\oplus M_{k(n)}$ is the quotient map. Since B sits where we want it to (by construction) the only question is whether or not B has any amenable traces. Unfortunately, it seems possible that B could have such traces. However, the following observation will keep our hopes alive.

LEMMA 6.5.5. *The image of B in the GNS representation of $\Pi M_{k(n)}/\oplus M_{k(n)}$ with respect to the trace τ_ω is isomorphic to $C_r^*(\Gamma)$ (which has no amenable traces).*

PROOF. Evidently we may identify the GNS representation of $\Pi M_{k(n)}/\oplus M_{k(n)}$, with respect to τ_ω, with a subalgebra (in fact, II_1-subfactor) of R^ω. Using this identification, it is clear that the resulting trace on $C^*(\Gamma)$ (which is just $\tau_\omega \circ \sigma \circ \pi$) is the canonical trace which vanishes on all non-trivial group elements. By uniqueness of GNS representations, the lemma follows. □

Thus we only need a result which states that amenable traces pass to quotients (for then B can't have any amenable traces since one of its quotients has no such traces). Unfortunately this is not true (e.g. the canonical trace on $C^*(\mathbb{F}_2)$ is amenable while it is not when regarded as a trace on $C_r^*(\mathbb{F}_2)$). Thus we will have to use a bit of trickery. The next observation gives us a way of forcing amenable traces to pass to quotients.

6.5. AMENABLE TRACES AND OBSTRUCTIONS IN K-HOMOLOGY

LEMMA 6.5.6. *Let X be a C^*-algebra, $J \subset X$ be an ideal, $J^\perp \subset X$ be the ideal perpendicular to J (i.e. $J^\perp = \{x \in X : xJ = Jx = 0\}$) and τ be an amenable trace on X. If $\tau|_{J^\perp} \neq 0$ then X/J has at least one amenable trace.*

PROOF. Let $\gamma = \frac{1}{\|\tau|_{J^\perp}\|}\tau|_{J^\perp}$ and Proposition 3.5.11 ensures that γ is an amenable trace on $J^\perp$. Since $J \cap J^\perp = \{0\}$ we may identify $J^\perp$ with an ideal in X/J. However, thanks to Proposition 3.5.12 we may extend the amenable trace γ to an amenable trace on X/J. □

We almost have all the pieces to the puzzle. We just have to find an appropriate ideal to add on to the algebra B above. To do this requires a bit more notation. Let $I_\omega \subset \Pi M_{k(n)}$ be the ideal of sequences (x_n) such that $\|x_n\|_2 \to 0$ as $n \to \omega$. Let $J = B \cap \sigma(I_\omega)$. Note that $B/J \cong C_r^*(\Gamma)$ by Lemma 6.5.5.

LEMMA 6.5.7. *(cf. [2]) There exists a projection $p \in \Pi M_{k(n)}/ \oplus M_{k(n)}$ and a unitary $u \in \Pi M_{k(n)}/ \oplus M_{k(n)}$ such that:*
  (1) $pJ = Jp = 0$.
  (2) $\tau_\omega(p) \geq 1/2$.
  (3) $p + upu^* \geq 1$.

Please see [2, Proposition, pg. 456] for the proof of this elementary (but crucial!) fact. (Actually, Anderson only constructs the projection and in [80] Wassermann makes the observation that the unitary u exists.) We now define the right algebras. Put
$$A = C^*(B, p) \subset W = C^*(B, p, u).$$

COROLLARY 6.5.8. *(cf. [2], [80]) $Ext(A)$ and $Ext(W)$ are not groups. W is not quasidiagonal, though it is the image (in the Calkin algebra) of a quasidiagonal set of operators on $B(H)$.*

PROOF. First note that J is also an ideal in A (since p is orthogonal to J) and A/J contains a unital copy of $C_r^*(\Gamma) \cong B/J$. By the sequence of lemmas above we have the following chain of implications: $Ext(A)$ is a group $\implies \tau_\omega|_A$ is amenable $\implies A/J$ has an amenable trace (since τ_ω restricted to the ideal generated by p – which is orthogonal to J – does not vanish). However the last statement provides a contradiction since $C_r^*(\Gamma) \subset A/J$ has no amenable traces.

More generally, note that this argument shows that any amenable trace on A must vanish on p. This implies that W has no amenable traces whatsoever since the equation $p + upu^* \geq 1$ implies that no trace on W can vanish on p. Hence $Ext(W)$ is not a group and W is not quasidiagonal. □

REMARK 6.5.9. In the case that $\Gamma = \mathbb{F}_2$, the algebra A is (essentially) the example given in [2] and the algebra W is the one given in [80].

We now turn to the property T case. We will need the following structure theorem for full group C*-algebras of property T groups.

THEOREM 6.5.10. *(cf. [41, 3.7.6]) Let Γ be a discrete group with Kazhdan's property T. Then for each finite dimensional, irreducible, unitary representation $\pi : \Gamma \to B(H_\pi)$ there is a central projection $p_\pi \in C^*(\Gamma)$ with the following properties:*
  (1) $p_{\pi_1} = p_{\pi_2}$ *if and only if π_1 and π_2 are unitarily equivalent and $p_{\pi_1} \perp p_{\pi_2}$ otherwise.*
  (2) $p_\pi C^*(\Gamma) = p_\pi C^*(\Gamma) p_\pi \cong B(H_\pi) = M_n$, *where $n = dim(\pi)$.*

(3) If $\rho\colon C^*(\Gamma) \to B(K)$ is a representation which contains the finite dimensional, irreducible representation π as a subrepresentation then $\rho(p_\pi) \neq 0$ (actually, $\rho(p_\pi)$ acts as the orthogonal projection onto the π-isotypical subspace).

With this result in hand, it is an easy matter to construct a C*-algebra with traces which are not amenable. Indeed, let $J \subset C^*(\Gamma)$ be the ideal generated by all the (central) projections coming from finite dimensional, irreducible representations. Note that $J \cong \oplus M_{k(n)}$ (possibly a finite direct sum) for some integers $k(n)$. Since the multiplier algebra of $\oplus M_{k(n)}$ is just $\Pi M_{k(n)}$ we get a ∗-homomorphism $\pi : C^*(\Gamma) \to \Pi M_{k(n)}$ such that $\pi(J) = \oplus M_{k(n)}$. Identifying $\Pi M_{k(n)}$ with the block diagonal operators on $H = \oplus \mathbb{C}^{k(n)}$ and letting $\sigma : B(H) \to Q(H)$ be the quotient map onto the Calkin algebra we define the right C*-algebra as:

$$W_\Gamma = \sigma \circ \pi(C^*(\Gamma)) \subset \Pi M_{k(n)} / \oplus M_{k(n)}.$$

COROLLARY 6.5.11. *(cf. [79]) If Γ is a discrete group with Kazhdan's property T and Γ has infinitely many non-equivalent, finite dimensional, irreducible representations (e.g. $SL(3,\mathbb{Z})$) then $\text{Ext}(W_\Gamma)$ is not a group and W_Γ is not quasidiagonal.*

PROOF. It suffices to show that W_Γ has no amenable traces. So assume that τ is amenable on W_Γ. Representing W_Γ faithfully on some Hilbert space K (i.e. $W_\Gamma \subset B(K)$) we can, by part (3) of Theorem 3.1.7, find unit vectors $v_n \in HS(K)$ (where $HS(K)$ is the Hilbert space of Hilbert-Schmidt operators on K) which are asymptotically invariant under the (conjugation) action of the unitary group of W_Γ on $HS(K)$.[11] Since Γ has property T it follows that this conjugation action on $HS(K)$ must have a fixed point. But this is precisely the same thing as saying that there is a Hilbert-Schmidt operator in the commutant W_Γ' and once you get a compact in the commutant you must also have a finite rank (spectral) projection in the commutant of W_Γ as well. But commuting finite rank projections give finite dimensional representations and this will give our contradiction. Indeed, we constructed W_Γ by first factoring out all of the irreducible finite dimensional representations from $C^*(\Gamma)$ (which implies that we factored out all finite dimensional representations as well). Hence, W_Γ can't have any finite dimensional representations since these would induce finite dimensional representations on $C^*(\Gamma)$ which factorize through W_Γ. □

REMARK 6.5.12. In [45] Kirchberg proved that a C*-algebra A has the local lifting property if and only if $\text{Ext}(S(A))$ is a group, where $S(A)$ denotes the unitization of $C_0(\mathbb{R}) \otimes A$. Ozawa has pointed out to us that the class of C*-algebras with the local lifting property is, in some sense, small and hence Kirchberg's result shows that Ext is not a group for lots of C*-algebras. (For example, 'most' finite dimensional operator spaces will generate C*-algebras without the local lifting property since the space of n-dimensional operator spaces is not separable – cf. [43] – while every finite dimensional subspace of a C*-algebra with the local lifting property can be identified with a subspace of $C^*(\mathbb{F}_\infty)$ and the space of n-dimensional operator subspaces of any separable C*-algebra is a separable set.)

---

[11] Since the Hilbert-Schmidt norm is invariant under left and right multiplication by unitaries, Connes' Følner condition is equivalent to $\|uv_n u^* - v_n\|_{HS} \to 0$, where $v_n = \frac{1}{\|P_n\|_{HS}} P_n$, whenever u is unitary.

However, it is interesting to note that all of the *concrete* examples of C*-algebras without the local lifting property fail to have this property because of invariant mean considerations (we are referring to concrete *separable* examples – $B(H)$ does not have the LLP, by [**43**]). Indeed, the only examples of C*-algebras (that we are aware of) without the local lifting property are the reduced group C*-algebras of non-amenable, residually finite discrete groups or groups which contain a subgroup isomorphic to such a group. Actually the most general statement we know is: If Γ contains a non-amenable subgroup H such that $C_r^*(H)$ embeds into R^ω then $C_r^*(\Gamma)$ does not have the LLP and hence $\mathrm{Ext}(S(C_r^*(\Gamma)))$ is not a group. (Proof: Since $C_r^*(H)$ has no amenable traces, yet it does embed into R^ω it follows that $C_r^*(H)$ does not have the LLP. Since there is a conditional expectation $C_r^*(\Gamma) \to C_r^*(H)$ it follows that $C_r^*(\Gamma)$ does not have the LLP either – cf. [**45**, Corollary 2.6.(v) and Proposition 3.1].)

6.6. Stable finiteness versus quasidiagonality

We now observe that approximation properties of traces are related to some important questions about quasidiagonal C*-algebras. At the moment there are only two known obstructions to quasidiagonality; (1) the existence of an infinite projection (in some matrix algebra over the given algebra) and (2) the absence of a quasidiagonal trace. In other words, every quasidiagonal C*-algebra is stably finite and has a quasidiagonal tracial state (recall that the latter statement is not true in the non-unital case). In [**78**] Voiculescu posed the challenging problem of finding a complete set of obstructions – preferably of a topological nature – to quasidiagonality. This ambitious goal is still out of our reach as the class of quasidiagonal C*-algebras is deceptively difficult to get a handle on. In this section we wish to describe two of the major open questions around quasidiagonality and point out the role that approximation properties of traces have already played as well as the role they may play in the future.

The two questions we will address are:

(1) Is every nuclear, stably finite C*-algebra necessarily quasidiagonal?
(2) Is the hyperfinite II$_1$-factor quasidiagonal?

The first question was asked by Blackadar and Kirchberg in [**13**]. This question is of fundamental importance in Elliott's classification program. Indeed, suppose that one could even show the weaker statement that every unital, simple, nuclear, stably finite C*-algebra with real rank zero was quasidiagonal. Then by Popa's result it would follow that every such algebra was a Popa algebra. Hence every such algebra would satisfy an internal finite dimensional approximation property which is similar to that which defines Lin's tracially AF algebras. In other words, question (1) above is 'half way' to completing the simple, stably finite, real rank zero case of Elliott's conjecture (the other half being a proof that nuclear, Popa algebras with real rank zero are tracially AF and hence fall under Lin's classification theorem). Clearly weaker variations of question (1) would already be of great interest. For example, what if one assumes a faithful trace? How about just the simple, real rank zero case?

Note also that question (1), or even the weaker version where one assumes a faithful trace, would immediately imply Rosenberg's conjecture that the reduced group C*-algebra of any discrete, amenable group is quasidiagonal. The second question above would also imply Rosenberg's conjecture as well as yield the faithful trace case of question (1) since every finite, hyperfinite von Neumann algebra can be embed into R. (Note that it seems unlikely that question (2) would imply question (1) in general since there are nuclear, stably finite – even quasidiagonal – C*-algebras which can't be embed into R since they have no *faithful* trace.)

In our opinion, it is far from clear which way either of the questions above will go. Suppose one wanted to give negative answers. One natural strategy for doing this would be to embed a C*-algebra which was not quasidiagonal into R or into some nuclear, stably finite C*-algebra (since quasidiagonality passes to subalgebras, this would give a contradiction). However, amenable trace considerations show that we cannot realize this strategy with our current set of examples.

PROPOSITION 6.6.1. *Let A be any C*-algebra which is known to not be quasidiagonal. Then it is impossible to embed A into R or into any nuclear, stably finite C*-algebra.*

PROOF. We caution the reader that there are plenty of stably finite C*-algebras for which it is not known whether they are quasidiagonal (cf. Rosenberg's Conjecture), but we now describe all the examples which are known to not be quasidiagonal.

The basic examples of non-quasidiagonal C*-algebras are: (1) Anything with an infinite projection (e.g. Cuntz algebras), (2) $C_r^*(\Gamma)$, where Γ is a discrete, non-amenable group and (3) the algebras constructed by Wassermann and denoted by W and W_Γ in the previous section of these notes. Since quasidiagonality passes to subalgebras, anything containing one of the examples above will also not be quasidiagonal (e.g. most reduced free products are not quasidiagonal since these tend to either be purely infinite or contain a copy of $C_r^*(\mathbb{F}_2)$). However, since the proposition at hand regards embeddings it will be sufficient to observe that none of the three basic examples above can be embed into R or into a nuclear, stably finite C*-algebra.

Excluding case (1) is trivial since stably finite C*-algebras never contain infinite projections. Excluding the reduced group C*-algebras of non-amenable, discrete groups has already been observed in Corollary 4.2.4 (recall the proof: $C_r^*(\Gamma)$ has no amenable traces, while R and any stably finite, nuclear C*-algebra does have amenable traces). Finally, Wassermann's examples also can't be embed into anything with an amenable trace because it was precisely the absence of such traces which allowed us to deduce that they were not quasidiagonal C*-algebras. □

REMARK 6.6.2. We wish to emphasize that the result above does not imply that one could not answer the questions above negatively by embedding a non-quasidiagonal C*-algebra into one of the algebras in question. It only says that we can't use any of the examples which are currently known to not be quasidiagonal. Of course, this takes us back to Voiculescu's original problem: what is a complete set of obstructions to quasidiagonality? Constructing new examples of non-quasidiagonal algebras would presumably require gaining insight to Voiculescu's question.

We should mention that several experts we have talked to feel that it is very unlikely that R is quasidiagonal. Their intuition seems to be based on the following fact which is known to many, though we were unable to find it written down anywhere. We thank George Elliott for showing us a very nice proof. We have only modified the last few lines of Elliott's argument so that we can deduce a slightly stronger statement.

LEMMA 6.6.3. *Let R act on $L^2(R)$ via the GNS construction. There is no sequence of nonzero, finite rank projections $P_1, P_2, \ldots$ such that $\|[x, P_n]\| \to 0$ for all $x \in R$.*

PROOF. The proof goes by contradiction. So let $P_1, P_2, \ldots$ be finite rank projections such that $\|[x, P_n]\| \to 0$ for all $x \in R$. Put $K = \oplus_{n \in \mathbb{N}} L^2(R) = L^2(R) \otimes_2 l^2(\mathbb{N})$ and $P = \oplus_{n \in \mathbb{N}} P_n$. Then $(x \otimes 1)P - P(x \otimes 1)$ is a compact operator for every $x \in R$. Hence, down in the Calkin algebra P will land in the commutant of $R \otimes 1$. But then by a theorem of Johnson and Parrott (see the remarks after [**42**, Lemma 3.3]) it follows that P is a compact perturbation of an element in the commutant of $R \otimes 1$. That is, there exists an infinite matrix $T = (T_{i,j})_{i,j \in \mathbb{N}}$ such that each $T_{i,j} \in R' \subset B(L^2(R))$ and $P - T$ is compact on K. In particular, this implies that $\|P_n - T_{n,n}\| \to 0$. Thus $\|T_{n,n}\| \to 1$ and down in the Calkin algebra the norm of $T_{n,n}$ is tending to zero. However this is a contradiction since the commutant of R is a II$_1$-factor (isomorphic to R) and hence a simple C*-algebra. Thus the mapping to the Calkin algebra is isometric. □

Note that the proof above never used the fact that the P_n's are projections and hence also holds for *sequences* of finite rank operators whose norms are tending to one. However, since we can always construct a quasicentral net of finite rank operators for R we are left to conclude that the lemma above has more to do with sequences versus nets (i.e. separable versus non-separable Hilbert spaces) than it does with quasidiagonality.

The point of the above discussion is that providing negative answers to either of the questions posed seems difficult at this point. The theory of amenable traces tell us that the embedding strategy is (currently!) hopeless, while the W*-obstruction appears to be less related to quasidiagonality issues than had been expected. As mentioned at the beginning, the only other obvious strategy is to prove that there exists a stably finite, nuclear C*-algebra with no quasidiagonal traces as this would certainly imply that such an example was not a quasidiagonal C*-algebra (recall, however, that all traces on nuclear C*-algebras are uniform amenable) or to show that the unique trace on R is not quasidiagonal. However we are not aware of a single example of an amenable trace which is not quasidiagonal. We believe such things exist, but constructing one seems quite difficult (note the parallel with Voiculescu's question). Our next result shows that uniform amenable traces may help unlock the quasidiagonality question for R.

PROPOSITION 6.6.4. *R is quasidiagonal if and only if for every (separable) C*-algebra A we have $\mathrm{AT}(A)_{\mathrm{QD}} \supset \mathrm{UAT}(A)$.*

PROOF. We begin with the necessity. Let A be arbitrary. It suffices to show that the extreme points of UAT(A) belong to AT(A)$_{\mathrm{QD}}$. However, every extreme point of UAT(A) is also an extreme point of T(A), since UAT(A) is a face, and hence gives R in the GNS representation. Thus, if we assume that R is quasidiagonal then its unique trace must belong to T$(R)_{\mathrm{QD}}$ and this completes the proof. (One may worry about non-separability issues here, but everything works fine with nets.)

For the sufficiency, we first point out that R is quasidiagonal if and only if all of its separable C*-subalgebras are quasidiagonal. So let $A \subset R$ be an arbitrary separable, unital subalgebra. Let $\tau \in \mathrm{T}(A)$ be the restriction of the unique trace on R to A. Clearly τ is faithful and belongs to UAT(A). Hence it also belongs to AT(A)$_{\mathrm{QD}}$. But we already saw in the proof of Proposition 4.1.3 that the existence of a faithful quasidiagonal trace implies C*-quasidiagonality of the algebra and hence the proof is complete. □

Applying the proposition above and Theorem 4.3.3 we immediately get the following corollary. Note, in particular, that if R is quasidiagonal then if would follow that every trace on a nuclear C*-algebra is uniform quasidiagonal (since we already know every trace on such an algebra is amenable).

COROLLARY 6.6.5. *If R is quasidiagonal then for every locally reflexive C*-algebra A we have*
$$\mathrm{AT}(A) = \mathrm{UAT}(A) = \mathrm{AT}(A)_{\mathrm{QD}} = \mathrm{UAT}(A)_{\mathrm{QD}}.$$

6.7. Questions

We wish to end these notes with a series of questions. Preliminary versions of this paper appeared in 2001 and hence some of the questions have been answered. However, we list all of our original questions just for completeness.

(1) Is every II_1-factor representation of a Popa algebra McDuff?[12]
(2) Is there an example such that $\mathrm{AT}(A) \neq \mathrm{AT}(A)_{\mathrm{QD}}$? How about $\mathrm{UAT}(A) \neq \mathrm{UAT}(A)_{\mathrm{QD}}$? The only obvious obstruction is related to quasidiagonality since the existence of a *faithful* trace in $\mathrm{AT}(A)_{\mathrm{QD}}$ implies quasidiagonality. However we just saw in the previous section that $\mathrm{AT}(A) = \emptyset$ for any of the standard examples of stably finite, non-quasidiagonal C*-algebras. For example, can one construct a C*-algebra which has the WEP and a *faithful* trace but which is not quasidiagonal? The most natural candidate is the hyperfinite II_1-factor.
(3) Can a free group or property T II_1-factor contain a weakly dense Popa algebra?[13]
(4) Can one give estimates of the free entropy dimension of a finite set of elements in a Popa algebra which is independent of the particular trace? This is related to the semicontinuity/invariance problem for free entropy dimension. Constructing a counterexample may be more reasonable and this would also be of interest.
(5) Can one prove a classification theorem for simple, nuclear, real rank zero C*-algebras which satisfy the UCT and such that $\mathrm{T}(A) = \mathrm{AT}(A)_{\mathrm{QD}}$ (= $\mathrm{UAT}(A)_{\mathrm{QD}}$, by Theorem 4.3.3)? In [**67**, Theorem 3.3] Popa proves that such algebras have an internal finite dimensional approximation property which should be of use. Presumably the role of nuclearity needs to be clarified as Popa never assumes nuclearity in [**67**].
(6) If A is locally reflexive (or even nuclear) do we always have that
$$\mathrm{UAT}(A)_{\mathrm{QD}} \cap \mathrm{AT}(A)_{\mathrm{LFD}} = \mathrm{UAT}(A)_{\mathrm{LFD}}?$$
Feel free to add on as many additional assumptions, such as real rank zero or stable rank one, as are necessary! For example, if A is a nuclear Popa algebra with real rank zero, stable rank one, unperforated K-theory and

---

[12] This question has been resolved in joint work with Ken Dykema: $L(\mathbb{F}_2)$, which is not McDuff, contains a weakly dense Popa algebra [**19**].

[13] Free group factors – at least on finitely many generators – do contain weakly dense Popa algebras [**19**], but the $L(\mathbb{F}_\infty)$ and property T cases are still open.

unique trace τ, is it necessarily true that τ is uniform locally finite dimensional? (In this setting we do know that $\tau \in \mathrm{UAT}(A)_{\mathrm{QD}} \cap \mathrm{AT}(A)_{\mathrm{LFD}}$.) See Proposition 6.1.23 for related questions.

(7) Can an infinite, simple, discrete group with Kazhdan's property T be embed into the unitary group of an R^ω-embeddable factor? (Compare with [**69**] where it is shown that no such embedding exists into the unitary group of $L(\mathbb{F}_n)$ or, more generally, $L(\Gamma)$ for any a-T-menable discrete group Γ.)

(8) Let Γ be a discrete group such that the group von Neumann algebra of Γ has the weak expectation property relative to a weakly dense C*-subalgebra (i.e. $L(\Gamma) \subset R^\omega$). Does it follow that Γ is an exact group? Perhaps just uniformly embeddable into Hilbert space? An affirmative answer would have three important consequences: (a) the Novikov conjecture is true for all residually finite groups (and every other group which embeds into R^ω) (b) counterexamples to Connes' embedding problem (hence a negative answer to Voiculescu's 'unification problem') since Gromov has constructed discrete groups which can't be uniformly embed into Hilbert space and (3) there exist hyperbolic groups which are not residually finite (since Gromov's examples are apparently inductive limits of hyperbolic groups and it is easy to see that inductive limits of groups which embed into R^ω are also embeddable into R^ω). For some time we felt confident that an affirmative answer was not far off. However, we have been unable to fill some gaps in our original strategy and the statement "R^ω-embeddable implies uniformly embeddable" has been downgraded to a conjecture. Recently Guentner, Higson and Weinberger showed that every linear group is exact and, hence, uniformly embeddable into Hilbert space (cf. [**32**]). Is there an extremely clever choice of microstates together with an ultraproduct argument which will work here?

(9) Is the canonical trace on $C^*(\Gamma)$ (giving the left regular representation) always an amenable trace whenever Γ is a hyperbolic group? Is every hyperbolic group embeddable into the unitary group of R^ω? An affirmative answer would imply that our previous question has a negative answer (since Gromov's non-exact groups are inductive limits of hyperbolic groups and hence would have to embed into R^ω) while a negative answer would imply that not every hyperbolic group is residually finite (since the canonical trace on $C^*(\Gamma)$ is amenable for every residually finite group).

Bibliography

1. R.C. Alperin, *An elementary account of Selberg's lemma*, Enseign. Math. **33** (1987), 269 - 273.
2. J. Anderson, *A C^*-algebra A for which Ext(A) is not a group*, Ann. of Math. **107** (1978), 455 - 458.
3. B. Arveson, *Notes on extensions of C^*-algebras*, Duke Math. J. **44** (1977), 329–355.
4. _____, *The role of C^*-algebras in infinite dimensional numerical linear algebra*, C^*-algebras: 1943 - 1993 (San Antonio, TX, 1993), 114–129, Contemp. Math. **167**, Amer. Math. Soc., Providence, RI, 1994.
5. _____, *C^*-algebras and numerical linear algebra*, J. Funct. Anal. **122** (1994), 333–360.
6. E. Bédos, *Notes on hypertraces and C^*-algebras*, J. Operator Thy. **34** (1995), 285–306.
7. _____, *On Følner nets, Szegö's theorem and other eigenvalue distribution theorems*, Exposition. Math. **15** (1997), 193–228.
8. M.E.B. Bekka, *Amenable unitary representations of locally compact groups*, Invent. Math. **100** (1990), 383–401.
9. B. Blackadar, *Weak expectations and injectivity in operator algebras*, Proc. Amer. Math. Soc. **68** (1978), 49–53.
10. _____, *Weak expectations and nuclear C^*-algebras*, Indiana Univ. Math. J. **27** (1978), 1021–1026.
11. _____, *K-theory for operator algebras*, Springer–Verlag, 1986.
12. B. Blackadar, A. Kumjian and M. Rørdam *Approximately central matrix units and the structure of noncommutative tori*, K-Theory **6** (1992), 267–284.
13. B. Blackadar and E. Kirchberg, *Generalized inductive limits of finite-dimensional C^*-algebras*, Math. Ann. **307** (1997), 343–380.
14. B. Blackadar and E. Kirchberg, *Inner quasidiagonality and strong NF algebras*, Pacific J. Math. **198** (2001), 307 – 329.
15. A. Böttcher, *C^*-algebras in numerical analysis*, Irish Math. Soc. Bull. No. 45 (2000), 57 – 133.
16. L.G. Brown and G.K. Pedersen, *C^*-algebras of real rank zero*, J. Funct. Anal. **99** (1991), 131–149.
17. N.P. Brown, *On quasidiagonal C^*-algebras*, Operator algebras and applications, 19–64, Adv. Stud. Pure Math., 38, Math. Soc. Japan, Tokyo, 2004.
18. _____, *Herrero's approximation problem for quasidiagonal operators*, J. Funct. Anal. **186** (2001), 360–365.
19. N.P. Brown and K.J. Dykema, *Popa algebras in free group factors*, J. Reine Angew. Math. **573** (2004), 157–180. .
20. M.D. Choi and E. Effros, *Separable nuclear C^*-algebras and injectivity*, Duke Math. J. **43** (1976), 309–322.
21. _____, *The completely positive lifting problem for C^*-algebras*, Ann. Math. **104** (1976), 585–609.
22. _____, *Injectivity and operator spaces*, J. Functional Anal. **24** (1977), 156–209.
23. A. Connes, *Classification of injective factors: cases II_1, II_∞, III_λ, $\lambda \neq 1$*, Ann. Math. **104** (1976), 73–115.
24. _____, *Compact metric spaces, Fredholm modules, and hyperfiniteness*, Ergod. Th. & Dynam. Sys. **9** (1989), 207–220.
25. M.D. Dadarlat, *On the approximation of quasidiagonal C^*-algebras*, J. Funct. Anal. **167** (1999), 69–78.
26. _____, *Nonnuclear subalgebras of AF algebras*, Amer. J. Math. **122** (2000), 581-597.

27. K.R. Davidson, *C*-algebras by example*, Fields Institute Monographs 6, American Mathematical Society, Providence, RI, 1996.
28. E.G. Effros and U. Haagerup, *Lifting problems and local reflexivity for C*-algebras*, Duke Math. J. **52** (1985), 103–128.
29. S. Eilers and G.A. Elliott, *The Reisz property for the K_*-group of a C*-algebra of minimal stable and real rank*, C.R. Math. Acad. Sci. Soc. R. Can. **25** (2003), 108 – 113.
30. G.A. Elliott and G. Gong, *On the classification of C*-algebras of real rank zero II*, Ann. of Math. **144** (1996), 497–610.
31. G.A. Elliott and J. Villadsen, *Perforated ordered K_0-groups*, Canad. J. Math. **52** (2000), 1164–1191.
32. E. Guentner, N. Higson and S. Weinberger, *The Novikov conjecture for linear groups*, preprint 2003.
33. U. Haagerup, *A new proof of the equivalence of injectivity and hyperfiniteness for factors on a separable Hilbert space*, J. Funct. Anal. **62** (1985), 160 – 201.
34. _____, *Quasitraces on exact C*-algebras are traces*, unpublished manuscript, 1991.
35. _____, *Spectral decomposition of all operators in a II_1-factor which is embeddable in R^ω*, MSRI notes, 2001.
36. U. Haagerup and S. Thorbjørnsen, *Random matrices and K-theory for exact C*-algebras*, Doc. Math. **4** (1999), 341–450.
37. _____, *A new application of random matrices: $Ext(C_r^*(\mathbb{F}_2))$ is not a group*, preprint 2003.
38. U. Haagerup and C. Winsløw, *The Effros-Marechal topology in the space of von Neumann algebras II*, J. Funct. Anal. **171** (2000), 401–431.
39. D. Hadwin, *A noncommutative moment problem*, Proc. Amer. Math. Soc. **129** (2001), 1785–1791.
40. R. Hagen, S. Roch and B. Silbermann, *C*-algebras and numerical analysis*. Monographs and Textbooks in Pure and Applied Mathematics, 236. Marcel Dekker, Inc., New York, 2001.
41. N. Higson and J. Roe, *Analytic K-homology*, Oxford Mathematical Monographs. Oxford Science Publication. Oxford University Press, Oxford, 2000.
42. B.E. Johnson and S.K. Parrott, *Operators commuting with a von Neumann algebra modulo the set of compact operators*, J. Funct. Anal. **11** (1972), 39–61.
43. M. Junge and G. Pisier, *Bilinear forms on exact operator spaces and $B(H) \otimes B(H)$*, Geom. Funct. Anal. **5** (1995), 329 – 363.
44. R.V. Kadison and J.R. Ringrose, *Fundamentals of the theory of operator algebras*, Vol. 2. Advanced Theory. Graduate Studies in Mathematics, 16. American Mathematical Society, Providence, RI, 1997.
45. E. Kirchberg, *On nonsemisplit extensions, tensor products and exactness of group C*-algebras*, Invent. Math. **112** (1993), 449–489.
46. _____, *Commutants of unitaries in UHF algebras and functorial properties of exactness*, J. reine angew. Math. **452** (1994), 39–77.
47. _____, *Discrete groups with Kazhdan's property T and factorization property are residually finite*, Math. Ann. **299** (1994), 551–563.
48. E. Kirchberg and W. Winter, *Covering dimension and quasidiagonality*, Internat. J. Math. **15** (2004), 63–85.
49. E.C. Lance, *On nuclear C*-algebras*, J. Funct. Anal. **12** (1973), 157–176.
50. H. Lin, *Classification of simple C*-algebras of tracial topological rank zero*, Duke Math. J. **125** (2004), 91–119.
51. _____, *Tracially AF C*-algebras*, Trans. Amer. Math. Soc. **353** (2001), 693–722.
52. _____, *The tracial topological rank of C*-algebras*, Proc. London Math. Soc.(3) **83** (2001), 199–234.
53. _____, *Traces and simple C*-algebras with tracial topological rank zero*, J. Reine Angew. Math. **568** (2004), 99 – 137.
54. Q. Lin and N.C. Phillips, *C*-algebras of minimal diffeomorphisms*, preprint.
55. S. Narayan, *Quasidiagonality of direct sums of weighted shifts*, Trans. Amer. Math. Soc. **332** (1992), 757 – 774.
56. A. Olshanskii, *SQ-universality of hyperbolic groups*, Sb. Math. **186** (1995), 1199 – 1211.
57. N. Ozawa, *Homotopy invariance of AF embeddability*, Geom. Funct. Anal. **13** (2003), 216–222.
58. _____, *About the QWEP conjecture*, Internat. J. Math. **15** (2004), 501–530.

59. _____, *There is no separable universal II_1-factor*, Proc. Amer. Math. Soc. **132** (2004), 487–490.
60. V. Paulsen, *Completely bounded maps and dilations*, Pitman Research Notes in Mathematics, vol. 146, Longman, 1986.
61. G. Pedersen, *C^*-algebras and their automorphism groups*, Academic Press, London, 1979.
62. N.C. Phillips, *Cancellation and stable rank for direct limits of recursive subhomogeneous algebras*, preprint 2001.
63. _____, *Real rank and property (SP) for direct limits of recursive subhomogeneous algebras*, preprint 2004.
64. _____, *Examples of nonisomorphic nonnuclear simple stably finite C^*-algebras with the same Elliott invariants*, manuscript in preparation.
65. S. Popa, *A short proof of "injectivity implies hyperfiniteness" for finite von Neumann algebras*, J. Operator Theory **16** (1986), 261–272.
66. _____, *On amenability in type II_1 factors*, Operator algebras and applications, Vol. 2, 173–183, London Math. Soc. Lecture Notes Ser., 136, Cambridge Univ. Press, Cambridge, 1988.
67. _____, *On local finite-dimensional approximation of C^*-algebras*, Pacific J. Math. **181** (1997), 141–158.
68. _____, *On a class of type II_1-factors with Betti numbers invariants*, preprint 2002.
69. A.G. Robertson, *Property (T) for II_1 factors and unitary representations of Kazhdan groups*, Math. Ann. **296** (1993), 547–555.
70. M. Rørdam, *On the structure of simple C^*-algebras tensored with a UHF-algebra*, J. Funct. Anal. **100** (1991), 1–17.
71. _____, *On the structure of simple C^*-algebras tensored with a UHF-algebra II*, J. Funct. Anal. **107** (1992), 255–269.
72. _____, *A simple C^*-algebra with a finite and an infinite projection*, Acta Math. **191** (2003), 109–142.
73. J. Rosenberg and C. Schochet, *The Künneth theorem and the universal coefficient theorem for Kasparov's generalized K-functor*, Duke Math. J. **55** (1987), 431–474.
74. R. Smucker, *Quasidiagonal weighted shifts*, Pacific J. Math. **98** (1982), 173–182.
75. A. Toms, *On the independence of K-theory and stable rank for simple C^*-algebras*, J. Reine Angew. Math. **578** (2005), 185–199. (to appear).
76. D.-V. Voiculescu, *On the existence of quasicentral approximate units relative to normed ideal. Part I*, J. Funct. Anal. **91** (1990), 1–36.
77. _____, *A note on quasi-diagonal C^*-algebras and homotopy*, Duke Math. J. **62** (1991), 267–271.
78. _____, *Around quasidiagonal operators*, Integr. Equ. and Op. Thy. **17** (1993), 137–149.
79. S. Wassermann, *C^*-algebras associated with groups with Kazhdan's property T*, Ann. of Math. **134** (1991), 423–431.
80. _____, *A separable quasidiagonal C^*-algebra with a nonquasidiagonal quotient by the compact operators*, Math. Proc. Cambridge Philos. Soc. **110** (1991), 143–145.
81. _____, *Exact C^*-algebras and related topics*, Lecture Notes Series, 19. Seoul National University, Seoul, 1994.
82. W. Winter, *On topologically finite dimensional simple C^*-algebras*, Math. Ann. **332** (2005), 843–878.

Editorial Information

To be published in the *Memoirs*, a paper must be correct, new, nontrivial, and significant. Further, it must be well written and of interest to a substantial number of mathematicians. Piecemeal results, such as an inconclusive step toward an unproved major theorem or a minor variation on a known result, are in general not acceptable for publication. Papers appearing in *Memoirs* are generally at least 80 and not more than 200 published pages in length. Papers less than 80 or more than 200 published pages require the approval of the Managing Editor of the Transactions/Memoirs Editorial Board.

As of July 31, 2006, the backlog for this journal was approximately 11 volumes. This estimate is the result of dividing the number of manuscripts for this journal in the Providence office that have not yet gone to the printer on the above date by the average number of monographs per volume over the previous twelve months, reduced by the number of volumes published in four months (the time necessary for preparing a volume for the printer). (There are 6 volumes per year, each containing at least 4 numbers.)

A Consent to Publish and Copyright Agreement is required before a paper will be published in the *Memoirs*. After a paper is accepted for publication, the Providence office will send a Consent to Publish and Copyright Agreement to all authors of the paper. By submitting a paper to the *Memoirs*, authors certify that the results have not been submitted to nor are they under consideration for publication by another journal, conference proceedings, or similar publication.

Information for Authors

Memoirs are printed from camera copy fully prepared by the author. This means that the finished book will look exactly like the copy submitted.

The paper must contain a *descriptive title* and an *abstract* that summarizes the article in language suitable for workers in the general field (algebra, analysis, etc.). The *descriptive title* should be short, but informative; useless or vague phrases such as "some remarks about" or "concerning" should be avoided. The *abstract* should be at least one complete sentence, and at most 300 words. Included with the footnotes to the paper should be the 2000 *Mathematics Subject Classification* representing the primary and secondary subjects of the article. The classifications are accessible from www.ams.org/msc/. The list of classifications is also available in print starting with the 1999 annual index of *Mathematical Reviews*. The Mathematics Subject Classification footnote may be followed by a list of *key words and phrases* describing the subject matter of the article and taken from it. Journal abbreviations used in bibliographies are listed in the latest *Mathematical Reviews* annual index. The series abbreviations are also accessible from www.ams.org/publications/. To help in preparing and verifying references, the AMS offers MR Lookup, a Reference Tool for Linking, at www.ams.org/mrlookup/. When the manuscript is submitted, authors should supply the editor with electronic addresses if available. These will be printed after the postal address at the end of the article.

Electronically prepared manuscripts. The AMS encourages electronically prepared manuscripts, with a strong preference for $\mathcal{AMS}$-LaTeX. To this end, the Society has prepared $\mathcal{AMS}$-LaTeX author packages for each AMS publication. Author packages include instructions for preparing electronic manuscripts, the *AMS Author Handbook*, samples, and a style file that generates the particular design specifications of that publication series. Though $\mathcal{AMS}$-LaTeX is the highly preferred format of TeX, author packages are also available in $\mathcal{AMS}$-TeX.

Authors may retrieve an author package from e-MATH starting from www.ams.org/tex/ or via FTP to ftp.ams.org (login as anonymous, enter username as password, and type cd pub/author-info). The *AMS Author Handbook* and the *Instruction Manual* are available in PDF format following the author packages link from www.ams.org/tex/. The author package can also be obtained free of charge by sending email to `tech-support@ams.org` (Internet) or from the Publication Division, American Mathematical Society, 201 Charles St., Providence, RI 02904-2294, USA. When requesting an author package, please specify $\mathcal{AMS}$-LaTeX or $\mathcal{AMS}$-TeX and the publication in which your paper will appear. Please be sure to include your complete mailing address.

Sending electronic files. After acceptance, the source file(s) should be sent to the Providence office (this includes any TeX source file, any graphics files, and the DVI or PostScript file).

Before sending the source file, be sure you have proofread your paper carefully. The files you send must be the EXACT files used to generate the proof copy that was accepted for publication. For all publications, authors are required to send a printed copy of their paper, which exactly matches the copy approved for publication, along with any graphics that will appear in the paper.

TeX files may be submitted by email, FTP, or on diskette. The DVI file(s) and PostScript files should be submitted only by FTP or on diskette unless they are encoded properly to submit through email. (DVI files are binary and PostScript files tend to be very large.)

Electronically prepared manuscripts can be sent via email to `pub-submit@ams.org` (Internet). The subject line of the message should include the publication code to identify it as a Memoir. TeX source files, DVI files, and PostScript files can be transferred over the Internet by FTP to the Internet node `e-math.ams.org` (130.44.1.100).

Electronic graphics. Comprehensive instructions on preparing graphics are available at `www.ams.org/jourhtml/graphics.html`. A few of the major requirements are given here.

Submit files for graphics as EPS (Encapsulated PostScript) files. This includes graphics originated via a graphics application as well as scanned photographs or other computer-generated images. If this is not possible, TIFF files are acceptable as long as they can be opened in Adobe Photoshop or Illustrator. No matter what method was used to produce the graphic, it is necessary to provide a paper copy to the AMS.

Authors using graphics packages for the creation of electronic art should also avoid the use of any lines thinner than 0.5 points in width. Many graphics packages allow the user to specify a "hairline" for a very thin line. Hairlines often look acceptable when proofed on a typical laser printer. However, when produced on a high-resolution laser imagesetter, hairlines become nearly invisible and will be lost entirely in the final printing process.

Screens should be set to values between 15% and 85%. Screens which fall outside of this range are too light or too dark to print correctly. Variations of screens within a graphic should be no less than 10%.

Inquiries. Any inquiries concerning a paper that has been accepted for publication should be sent directly to the Electronic Prepress Department, American Mathematical Society, 201 Charles St., Providence, RI 02904, USA.

Editors

This journal is designed particularly for long research papers, normally at least 80 pages in length, and groups of cognate papers in pure and applied mathematics. Papers intended for publication in the *Memoirs* should be addressed to one of the following editors. In principle the Memoirs welcomes electronic submissions, and some of the editors, those whose names appear below with an asterisk (*), have indicated that they prefer them. However, editors reserve the right to request hard copies after papers have been submitted electronically. Authors are advised to make preliminary email inquiries to editors about whether they are likely to be able to handle submissions in a particular electronic form.

*Algebra to ALEXANDER KLESHCHEV, Department of Mathematics, University of Oregon, Eugene, OR 97403-1222; email: ams@noether.uoregon.edu

Algebra and its application to MINA TEICHER, Emmy Noether Research Institute for Mathematics, Bar-Ilan University, Ramat-Gan 52900, Israel; email: teicher@macs.biu.ac.il

Algebraic geometry to DAN ABRAMOVICH, Department of Mathematics, Brown University, Box 1917, Providence, RI 02912; email: amsedit@math.brown.edu

*Algebraic number theory to V. KUMAR MURTY, Department of Mathematics, University of Toronto, 100 St. George Street, Toronto, ON M5S 1A1, Canada; email: murty@math.toronto.edu

*Algebraic topology to ALEJANDRO ADEM, Department of Mathematics, University of British Columbia, Room 121, 1984 Mathematics Road, Vancouver, British Columbia, Canada V6T 1Z2; email: adem@math.ubc.ca

*Combinatorics to JOHN R. STEMBRIDGE, Department of Mathematics, University of Michigan, Ann Arbor, Michigan 48109-1109; email: FRS@umich.edu

Complex analysis and harmonic analysis to ALEXANDER NAGEL, Department of Mathematics, University of Wisconsin, 480 Lincoln Drive, Madison, WI 53706-1313; email: nagel@math.wisc.edu

*Differential geometry and global analysis to LISA C. JEFFREY, Department of Mathematics, University of Toronto, 100 St. George St., Toronto, ON Canada M5S 3G3; email: jeffrey@math.toronto.edu

Dynamical systems and ergodic theory to AMIE WILKINSON, Department of Mathematics, Northwestern University, 2033 Sheridan Road, Evanston, IL 60208-2730; email: transactions@math.northwestern.edu

*Functional analysis and operator algebras to MARIUS DADARLAT, Department of Mathematics, Purdue University, 150 N. University St., West Lafayette, IN 47907-2067; email: mdd@math.purdue.edu

*Geometric analysis to TOBIAS COLDING, Courant Institute, New York University, 251 Mercer St., New York, NY 10012; email: traneditor@cims.nyu.edu

*Geometric analysis to MLADEN BESTVINA, Department of Mathematics, University of Utah, 155 South 1400 East, JWB 233, Salt Lake City, Utah 84112-0090; email: bestvina@math.utah.edu

Harmonic analysis, representation theory, and Lie theory to ROBERT J. STANTON, Department of Mathematics, The Ohio State University, 231 West 18th Avenue, Columbus, OH 43210-1174; email: stanton@math.ohio-state.edu

*Logic to STEFFEN LEMPP, Department of Mathematics, University of Wisconsin, 480 Lincoln Drive, Madison, Wisconsin 53706-1388; email: lempp@math.wisc.edu

*Ordinary differential equations, and applied mathematics to PETER W. BATES, Department of Mathematics, Michigan State University, East Lansing, MI 48824-1027; email: bates@math.msu.edu

*Partial differential equations to GUSTAVO PONCE, Department of Mathematics, South Hall, Room 6607, University of California, Santa Barbara, CA 93106; email: ponce@math.ucsb.edu

*Probability and statistics to KRZYSZTOF BURDZY, Department of Mathematics, University of Washington, Box 354350, Seattle, Washington 98195-4350; email: burdzy@math.washington.edu

*Real analysis and partial differential equations to DANIEL TATARU, Department of Mathematics, University of California, Berkeley, Berkeley, CA 94720; email: tataru@math.berkeley.edu

All other communications to the editors should be addressed to the Managing Editor, ROBERT GURALNICK, Department of Mathematics, University of Southern California, Los Angeles, CA 90089-1113; email: guralnic@math.usc.edu.

Titles in This Series

868 **Gelu Popescu,** Entropy and multivariable interpolation, 2006
867 **Vilmos Totik,** Metric properties of harmonic measures, 2006
866 **William Craig,** Semigroups underlying first-order logic, 2006
865 **Nathanial P. Brown,** Invariant means and finite representation theory of C^*-algebras, 2006
864 **John M. Lee,** Fredholm operators and Einstein metrics on conformally compact manifolds, 2006
863 **M. Lübke and A. Teleman,** The Universal Kobayashi-Hitchin correspondence on Hermitian manifolds, 2006
862 **Alberto Canonaco,** The Beilinson complex and canonical rings of irregular surfaces, 2006
861 **Leon A. Takhtajan and Lee-Peng Teo,** Weil-Petersson metric on the universal Teichmüller space, 2006
860 **Thomas M. Fiore,** Pseudo limits, biadjoints and pseudo algebras: Categorical foundations of conformal field theory, 2006
859 **N. Arcozzi, R. Rochberg, and E. Sawyer,** Carleson measures and interpolating sequences for Besov spaces on complex balls, 2006
858 **Enrico Valdinoci, Berardino Sciunzi, and Vasile Ovidiu Savin,** Flat level set regularity of p-Laplace phase transitions, 2006
857 **Donatella Danielli, Nocola Garofalo, and Duy-Minh Nhieu,** Non-doubling Ahlfors measures, perimeter measures, and the characterization of the trace spaces of Sobolev functions in Carnot-Carathéodory spaces, 2006
856 **Vladimir Bolotnikov and Harry Dym,** On boundary interpolation for matrix valued Schur functions, 2006
855 **Yevgenia Kashina, Yorck Sommerhäuser, and Yongchang Zhu,** On higher Frobenius-Schur indicators, 2006
854 **Noam Greenberg,** The role of true finiteness in the admissible recursively enumerable degrees, 2006
853 **Joachim Krieger,** Stability of spherically symmetric wave maps, 2006
852 **Viorel Barbu, Irena Lasiecka, and Roberto Triggiani,** Tangential boundary stabilization of Navier-Stokes equations, 2006
851 **Jie Wu,** On maps from loop suspensions to loop spaces and the shuffle relations on the Cohen groups, 2006
850 **Siegfried Echterhoff, S. Kaliszewski, John Quigg, and Iain Raeburn,** A categorical approach to imprimitivity theorems for C^*-dynamical systems, 2006
849 **Katsuhiko Kuribayashi, Mamoru Mimura, and Tetsu Nishimoto,** Twisted tensor products related to the cohomology of the classifying spaces of loop groups, 2006
848 **Bob Oliver,** Equivalences of classifying spaces completed at the prime two, 2006
847 **Eric T. Sawyer and Richard L. Wheeden,** Hölder continuity of weak solutions to subelliptic equations with rough coefficients, 2006
846 **Victor Beresnevich, Detta Dickinson, and Sanju Velani,** Measure theoretic laws for lim–sup sets, 2006
845 **Ehud Friedgut, Vojtech Rödl, Andrzej Ruciński, and Prasad V. Tetali,** A Sharp threshold for random graphs with a monochromatic triangle in every edge coloring, 2006
844 **Amadeu Delshams, Rafael de la Llave, and Tere M. Seara,** A geometric mechanism for diffusion in Hamiltonian systems overcoming the large gap problem: Heuristics and rigorous verification on a model, 2006
843 **Denis V. Osin,** Relatively hyperbolic groups: Intrinsic geometry, algebraic properties, and algorithmic problems, 2006
842 **David P. Blecher and Vrej Zarikian,** The calculus of one-sided M-ideals and multipliers in operator spaces, 2006

TITLES IN THIS SERIES

841 **Enrique Artal Bartolo, Pierrette Cassou-Noguès, Ignacio Luengo, and Alejandro Melle Hernández,** Quasi-ordinary power series and their zeta functions, 2005

840 **Sławomir Kołodziej,** The complex Monge-Ampère equation and pluripotential theory, 2005

839 **Mihai Ciucu,** A random tiling model for two dimensional electrostatics, 2005

838 **V. Jurdjevic,** Integrable Hamiltonian systems on complex Lie groups, 2005

837 **Joseph A. Ball and Victor Vinnikov,** Lax-Phillips scattering and conservative linear systems: A Cuntz-algebra multidimensional setting, 2005

836 **H. G. Dales and A. T.-M. Lau,** The second duals of Beurling algbras, 2005

835 **Kiyoshi Igusa,** Higher complex torsion and the framing principle, 2005

834 **Kenichi Ohshika,** Kleinian groups which are limits of geometrically finite groups, 2005

833 **Greg Hjorth and Alexander S. Kechris,** Rigidity theorems for actions of product groups and countable Borel equivalence relations, 2005

832 **Lee Klingler and Lawrence S. Levy,** Representation type of commutative Noetherian rings III: Global wildness and tameness, 2005

831 **K. R. Goodearl and F. Wehrung,** The complete dimension theory of partially ordered systems with equivalence and orthogonality, 2005

830 **Jason Fulman, Peter M. Neumann, and Cheryl E. Praeger,** A generating function approach to the enumeration of matrices in classical groups over finite fields, 2005

829 **S. G. Bobkov and B. Zegarlinski,** Entropy bounds and isoperimetry, 2005

828 **Joel Berman and Paweł M. Idziak,** Generative complexity in algebra, 2005

827 **Trevor A. Welsh,** Fermionic expressions for minimal model Virasoro characters, 2005

826 **Guy Métivier and Kevin Zumbrun,** Large viscous boundary layers for noncharacteristic nonlinear hyperbolic problems, 2005

825 **Yaozhong Hu,** Integral transformations and anticipative calculus for fractional Brownian motions, 2005

824 **Luen-Chau Li and Serge Parmentier,** On dynamical Poisson groupoids I, 2005

823 **Claus Mokler,** An analogue of a reductive algebraic monoid whose unit group is a Kac-Moody group, 2005

822 **Stefano Pigola, Marco Rigoli, and Alberto G. Setti,** Maximum principles on Riemannian manifolds and applications, 2005

821 **Nicole Bopp and Hubert Rubenthaler,** Local zeta functions attached to the minimal spherical series for a class of symmetric spaces, 2005

820 **Vadim A. Kaimanovich and Mikhail Lyubich,** Conformal and harmonic measures on laminations associated with rational maps, 2005

819 **F. Andreatta and E. Z. Goren,** Hilbert modular forms: Mod p and p-adic aspects, 2005

818 **Tom De Medts,** An algebraic structure for Moufang quadrangles, 2005

817 **Javier Fernández de Bobadilla,** Moduli spaces of polynomials in two variables, 2005

816 **Francis Clarke,** Necessary conditions in dynamic optimization, 2005

815 **Martin Bendersky and Donald M. Davis,** V_1-periodic homotopy groups of $SO(n)$, 2004

814 **Johannes Huebschmann,** Kähler spaces, nilpotent orbits, and singular reduction, 2004

813 **Jeff Groah and Blake Temple,** Shock-wave solutions of the Einstein equations with perfect fluid sources: Existence and consistency by a locally inertial Glimm scheme, 2004

For a complete list of titles in this series, visit the AMS Bookstore at **www.ams.org/bookstore/**.